SpringerBriefs in Earth Sciences

More information about this series at http://www.springer.com/series/8897

V.I. Osipov

Physicochemical Theory of Effective Stress in Soils

 Springer

V.I. Osipov
Sergeev Institute of Environmental
    Geoscience RAS
Moscow
Russia

The Work is a translation of the book in Russian by V.I. Osipov, "Physicochemical Theory of Effective Stress in Soils", Moscow, IFZ RAN, 2012, ISBN 978-5-91682-016-4.

ISSN 2191-5369   ISSN 2191-5377   (electronic)
SpringerBriefs in Earth Sciences
ISBN 978-3-319-20638-7   ISBN 978-3-319-20639-4   (eBook)
DOI 10.1007/978-3-319-20639-4

Library of Congress Control Number: 2015946070

Springer Cham Heidelberg New York Dordrecht London

Printed on acid-free paper

Springer International Publishing AG Switzerland is part of Springer Science+Business Media (www.springer.com)

Preface

The theory of effective stresses developed by K. Terzaghi in the early 1920s lays down the fundamentals of the modern soil mechanics. According to this theory, all changes in soils are related to external gravitational stress (σ') transmitted to the soil skeleton, i.e., the effective stress, which may be found from equation $\sigma' = \sigma - u$, where σ is the total stress and u is the neutral stress transmitted to the pore water. The Terzaghi theory is successfully applied to problems related to consolidation of porous permeable soils, sand liquefaction during earthquakes, as well as to a number of other tasks. At the same time, the practical experience shows that the Terzaghi theory cannot be adequately applied to low-permeable fine-grained soils (clay) with closed porosity, as it leads to the discrepancy between calculated and experimentally obtained data. The explanation is that this theory considers soil as a homogeneous unstructured body, and does not take into account the existence of internal forces of molecular, electrostatic, and structural mechanical origin manifested in thin hydrate films of adsorbed water molecules at the contacts of structural elements, which produce the so-called internal disjoining pressure. These forces can be estimated based on the theoretical achievements of molecular physics and colloidal chemistry.

The book presents a new theory of effective stresses in soils. Unlike the well-known Terzaghi theory, it considers soil as a structured system with acting external and internal stresses of gravitational and physicochemical nature. Stresses calculated with the consideration of both stress types correspond to the actual effective stresses existent in a structured porous body.

The book characterizes both types of stresses and provides equations for calculating the total actual effective stresses in soil. The calculation of the internal stresses is based on the theory of contact interactions and the theory of disjoining effect of boundary hydrate films. Modern colloid chemistry and molecular physics permit interpreting the theory of effective stresses in terms of physicochemical mechanics.

The theory presented in this book was first published in Russia in 2012. The current edition is based on the original Russian text of the book with minor revision. The book covers mainly the results of long-term investigations performed by

Russian experts in physical chemistry and molecular physics under the guidance of P.A. Rebinder and B.V. Deryaguin, Full Members of the Russian Academy of Sciences. Scientists from other countries have been also conducting studies in this field. Therefore, this novel theory demonstrates the general modern trend in the development of soil mechanics.

Contents

Introduction

The theory of effective stresses developed by K. Terzaghi in the early 1920s serves as the foundation of the modern soil and rock mechanics. This theory has been advanced further by A. Skempton, de Boer, J. Mitchell and K. Soga, and many others. K. Terzaghi wrote in one of his papers that stresses, at any point of a soil body, may be calculated from the *total main stresses* σ_1, σ_2, and σ_3, acting at this point. If soil pores are filled with water under pressure u, the total main stress consists of two components. One component, i.e., u, acts both in the water and the solid parts of soil in any direction with equal intensity. This component is referred to as the *neutral stress* (or pore pressure of water). The balance $\sigma_1' = \sigma_1 - u$, $\sigma_2' = \sigma_2 - u$, and $\sigma_3' = \sigma_3 - u$ represents the stress excess over the neutral stress u, which spreads only through the solid skeleton of soil. This portion of the total main stresses is called the *main effective stress*. Changes in the neutral stress u have virtually no effect on soil compaction and its stress for failure. Porous media, such as sand, clay, or cement, respond to variation in u as if they were incompressible bodies possessing no internal friction. All visible effects related to changing stress, such as compressibility, destruction, or changing shear strength, are caused solely by variation in the effective stresses σ_1', σ_2' and σ_3'. Hence, any study on water-saturated soil stability requires the knowledge about both types of stresses, i.e., effective and neutral.

The Terzaghi theory has been widely applied in practice. It serves as a cornerstone for solution of numerous problems in coarse clastic and sandy soils and fractured rocks. At the same time, the application of this theory to fine-grained clay often leads to discrepancy between the data obtained and theoretical calculations. The discrepancies increase with the growing clay content in soils and the decreasing degree of their compaction.

The existing difficulties are associated with the fact that the Terzaghi theory considers the total effective stresses in soil and does not take into account stress distribution in the contact area between structural elements. In particular, this theory does not consider:

1. characteristics of stress distribution at the contacts with various geometry and energy;
2. presence of thin films of bound water providing for disjoining pressure at the contacts;
3. development of different processes of the physicochemical origin at the contacts, which induces the development of internal stress;
4. dependence between pore pressure and the nature of pore space in clay.

At present, considerable success has been achieved in the study of contact interaction in fine-grained bodies. This field has been primarily developed by P.A. Rebinder, Full Member of the USSR Academy of Sciences, and his numerous followers, who launched a new interdisciplinary scientific branch, i.e., *physicochemical mechanics of porous bodies*, which has integrated the modern achievements in molecular physics, colloidal chemistry, and classic mechanics.

Investigation of the processes acting in fine-grained systems along with various surface forces is very important for the stress study. Academician B.V. Deryagin and his followers have fundamentally contributed to the study of this issue by developing *the theory of the disjoining effect produced by thin hydrate films*.

The achievements of these Russian scientific schools allowed us to further develop the theory of effective stresses in soils and to interpret this theory from the physicochemical principles.

References

Mitchell JK, Soga K (2005) Fundamentals of soil behavior, 3rd edn. Wiley, New York
Skempton AW (1960) Significance of Terzaghi's concept of effective stress. In: Bjerrum L, Casagrante A, Peck R, Skempton AW (eds) From theory to practice in soil mechanics. Wiley, New York

Chapter 1
Stresses in Soils

1.1 Types of Stress

The stress in soil is a distributed force applied to an infinitesimal soil unit of a certain direction. It is essential that two vectors are included in the stress definition, i.e., the force vector and the vector normal to the unit surface.

According to the ideas adopted in the continuum mechanics, all stresses acting at any point of a body may be subdivided into normal and tangential components. The sum of normal and tangential stresses defines the stress conditions at this point. The stresses acting normally to the three mutually orthogonal planes, where tangential stresses are absent, *are called the main stresses* σ_1, σ_2, and σ_3. The stress is characterized by force acting per unit area of the studied body (kG/cm^2, kN/m^2, $dyne/cm^2$, kPa, atm, or bar).

According to their origin, the total stresses developing in soil or rock systems are subdivided into external and internal stresses. External stresses may be either natural or human-induced. Natural stress produced by the weight of overlying ground layers is referred to *as geostatic stress* (σ_g). If the soil is below the groundwater table, *hydrostatic stress* (σ_h) controlled by the weight of the water column above acts in soil pores. In this case, the geostatic pressure is found by subtracting the specific weight of the displaced water volume (hydrostatic uplift pressure) from the specific weight of soil.

In the presence of a confined aquifer, the aquiclude is affected by *hydrodynamic* stresses due to the existing water head and groundwater filtration (σ_{hd}).

Human-induced stresses are transmitted from the engineering structures or result from other static and dynamic effects developing in the course of human activity (σ_t).

Internal stresses are caused by physicochemical processes developing inside soils and tending to either decrease or increase their internal energy. The resulting internal stresses thus relate to molecular, electrostatic, structural-mechanic ($\Pi(h)$), and capillary (σ_c) forces.

© The Author(s) 2015
V.I. Osipov, *Physicochemical Theory of Effective Stress in Soils*,
SpringerBriefs in Earth Sciences, DOI 10.1007/978-3-319-20639-4_1

Depending on soil moisture content, the total stresses may be either transmitted to the soil skeleton or distributed between the skeleton and water filling the soil pores. In the unsaturated state, the stresses are transmitted completely to the soil skeleton. When the pores are filled with water, the stress is distributed between the skeleton and the pore water. The stresses transmitted to the soil skeleton are called the *effective stresses* (σ'). The stresses transmitted to the pore water form the *pore pressure* (u). In open porous systems, where all pores are hydraulically connected, the pore pressure is hydrostatic and is controlled by the weight of the water column above. In this case, it is called the *hydrostatic or neutral* (u_0) pressure. The hydrostatic pressure is uniform and equal in all directions. It causes the hydraulic uplift of soil and remains constant under steady hydrogeological conditions. Therefore, it is often pointed out that the hydrostatic pressure does not affect the main effective stresses. That is why it is called the neutral pressure. This statement is valid only for the constant u_0 value. Otherwise, it has influence on the effective stresses. For example, it is well known that the lowering groundwater level causes a decrease in the hydrostatic uplift pressure and, as a result, an additional compaction of clay mass.

The effective stresses transmitted to the soil skeleton are concentrated in the contact zones and produce the contact stresses transferred from one structural unit to another. A contact area is much smaller than a conventional area (cm^2, m^2) used in calculation of the average total effective stresses. Therefore, the stress in the contact zones exceeds manifold the average total effective stresses.

Measurement of the contact stresses is associated with many difficulties due to a complicated determination of the contact area. This may be performed experimentally only by measuring disjoining pressure produced by hydrate films of bound water in the contact zones.

In the absence of pore pressure, the effective stresses are considered to be completely transmitted to the soil skeleton through the contact zones. This is true only for the direct contact between mineral particles (Fig. 1.1a). The pattern is different in clay systems. In clay of the initial, low, or even medium lithification degree, the contact stresses are transferred through the bound water films (Fig. 1.1b). Therefore, the effective stress is compensated by the disjoining effect of these films ($\Pi_{(h)}$) and is not transmitted to the soil skeleton. The soil skeleton receives stress at such contacts only if the particles squeeze the hydrate film out to form "dry" contacts.

Thus in clay systems, stress is transformed in the contact zones and only a portion of the total effective stress (σ') may be transmitted to the soil skeleton. Let us call this stress component *the actual total effective stress* (σ''). The remaining part of the total effective stress is compensated by the bound water films.

Thus, a number of different external and internal stresses of gravitational and physicochemical origin, which depend on various factors and follow specific development laws, act in soils. Generally, it may be stated that the actual total effective stress transmitted to the soil skeleton is a complex function. Considering the superposition principle for calculation of stresses in soil and assuming that forces of different origin act independently, the total actual effective stress is controlled by the following components:

$$\sigma'' = \sigma_g + \sigma_h + \sigma_{hd} + \sigma_t - u - \Pi_{(h)} + \sigma_c \qquad (1.1)$$

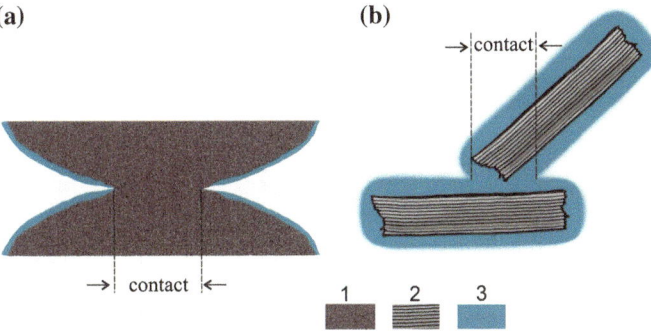

Fig. 1.1 Direct contact between sand grains (**a**) and clay particles (**b**): *1* sand grains; *2* clay particles, *3* bound water

The stresses in Eq. (1.1) may have different values or be asynchronous. For example, in water-saturated clay, there are no stresses caused by capillary forces, whereas, disjoining forces related to the developing adsorption and osmotic processes reach their maximum. The pattern may be different in non-saturated soil, where capillary forces play an important role along with adsorption and osmosis.

1.2 External Stresses

Stresses that originate in soil due to external impacts are called *external stresses*. These impacts include: (a) geostatic stress produced by the weight of overlying soil or rock strata; (b) hydrostatic and hydrodynamic stresses produced by the weight of water column in the water-saturated ground and by water filtration; and (c) stress produced by buildings and engineering structures. Though tectonic stress is also classified as external, its estimation is a complex problem. Therefore, engineering practice is limited to the assessment of stresses produced by the first three mentioned impacts.

1.2.1 Geostatic Stresses

Geostatic stress is produced by gravitation and is controlled by the ground weight above the examined point. For the horizontal day surface, normal geostatic stress increases with depth (z) as follows:

$$\sigma_{g(Z)} = \int_0^z \gamma_z \partial z, \qquad (1.2)$$

where γ_z is the specific weight of the overlying deposits together with the pore water.

For a soil body with a uniform density, the equation is simplified:

$$\sigma_{g(Z)} = \gamma z. \tag{1.3}$$

At a depth of about 100 m, this stress may reach 2–2.5 MPa.

For an elastic isotropic system without lateral deformations, the horizontal geostatic stresses $\sigma_{g(x)}$ and $\sigma_{g(y)}$ are calculated from the equation:

$$\sigma_{g(x)} = \sigma_{g(y)} = \xi \int_0^z \gamma_z \partial z, \tag{1.4}$$

where ξ is the lateral pressure coefficient tied with the Poisson coefficient (μ): $\xi = \mu/(1 - \mu)$.

In an unsaturated soil mass, all geostatic stresses are effective stresses, i.e., transmitted to the soil skeleton. For uniform soil density, the geostatic pressure is calculated from Eq. (1.3). An increase of geostatic pressure with depth can be presented graphically as a straight-line diagram (Fig. 1.2a). For a heterogeneous soil body, the geostatic pressure is calculated from Eq. (1.2). The calculations use the specific weight and the thickness of individual lithological and stratigraphic layers.

In water-saturated porous soils containing free water, the hydrostatic pressure u_0 should be considered, as it creates the uplift effect according to the Archimedean law:

$$\gamma' = (\gamma_s - \gamma_w)(1 - n), \tag{1.5}$$

where γ' is the specific weight of the hydraulically uplifted soil, γ_s is the specific weight of solid (mineral) soil particles, γ_w is the specific weight of water, and n is the porosity in unit fractions.

Thus, considering the hydrostatic uplift, we arrive at the effective geostatic stress created in the skeleton of water-saturated soil. In this case, the distribution

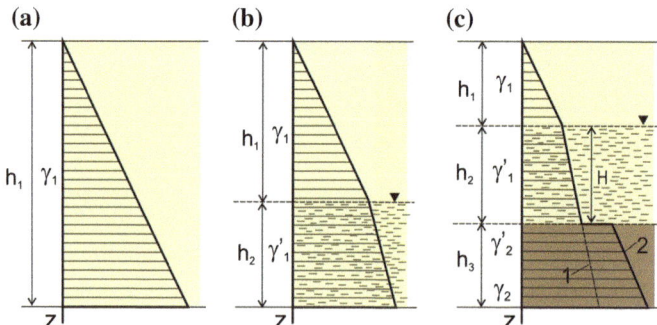

Fig. 1.2 Depth distribution of geostatic stress in: **a** an unsaturated soil body, **b** a soil body containing an unconfined aquifer, and **c** a soil body containing a clay horizon and an unconfined aquifer

diagram of the geostatic stress represents a straight line with a kink below the unconfined aquifer surface (Fig. 1.2b); the total stress in the soil body containing an unconfined aquifer is:

$$\sigma_{g(z)} = \gamma_1 h_1 + \gamma_1' h_2, \tag{1.6}$$

where γ_1' is the specific weight of hydraulically uplifted soil calculated from Eq. (1.5).

In clay soils, this is valid only for poorly lithified open clay systems, with free water contained in clay pores. Therefore, the layer of unlithified or poorly lithified clay is permeable and is in the hydraulically uplifted state (Fig. 1.2c, line 1).

The pattern is different in lithified unfractured clay masses. Such clay contains numerous closed pores with immobilized free or bound water. The clay acts as an aquitard, with no hydraulic connection between the pores, and, hence, with no hydraulic uplift. Not only the effective stress of the overlying strata is transmitted to the upper aquitard boundary, but also is the hydrostatic pressure of the water column weight, equal to:

$$u_0 = \gamma_w \, Hn, \tag{1.7}$$

where γ_w is the specific weight of water, H is the water column height, and n is the porosity of water-saturated soil.

Therefore, geostatic pressure rises quickly at the boundary with an aquiclude; whereas, within the clay layer, geostatic pressure increases according to the pattern typical of soils not subjected to hydraulic uplift (Fig. 1.2c, line 2).

Thus, the total geostatic stress in a soil mass with a clay layer at the base and a single unconfined aquifer is:

(a) permeable clay layer:

$$\sigma_{g(z)} = \gamma_1 h_1 + \gamma_1' h_2 + \gamma_2' h_3; \tag{1.8}$$

(b) impermeable clay layer:

$$\sigma_{g(z)} = \gamma_1 h_1 + \gamma_1' h_2 + \gamma_w Hn + \gamma_2 h_3. \tag{1.9}$$

If there are two aquifers (confined and unconfined), whose water columns have heights H_1 and H_2, respectively (Fig. 1.3), additional stresses appear and are controlled by the ratio between the actual (Ia) and initial (Ii) filtration gradients. Both gradients are found from:

$$I = \frac{H_1 - H_2}{h}, \tag{1.10}$$

where $H_1 - H_2$ is the difference between the hydraulic heads, and h is the permeable layer thickness.

The actually existing heads in the confined (H_1) and unconfined (H_2) aquifers are used for calculation of Ia.

Fig. 1.3 Depth distribution
of geostatic stress in a soil
body with two (confined and
unconfined) aquifers

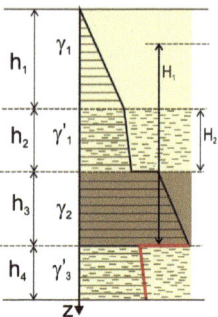

Ii value is found experimentally; H_1 and H_2 values are assigned, and the difference $H_1 - H_2$ is determined for the start of filtration. For $Ia \le Ii$, the clay layer is
regarded as an aquitard. Under these conditions, the aquitard roof is subjected not
only to the effective stress from the above-lying soil, but also to the weight of the
water column H_2, causing σ_z to increase by $\gamma_w H_2 n$ (Fig. 1.3). The stress from the
aquitard itself increases according to the product $\gamma_2 h_3$. However, σ_z falls abruptly
at the foot of the layer due to hydrostatic pressure $\gamma_w H_1$ from the existing water
head (H_1). Thus, the total geostatic pressure at the aquitard foot is:

$$\sigma_{g(z)} = \gamma_1 h_1 + \gamma_1' h_2 + \gamma_w H_2 n + \gamma_2 h_3 - \gamma_w H_1. \qquad (1.11)$$

Hydrostatic pressure produced by a confined aquifer may create a considerable
counter-pressure in soils, reaching several MPa. This causes compaction of soils
and subsidence of the earth surface when the groundwater head is relieved. This
phenomenon most often manifests itself during large-scale artesian water intake in
urban, industrial, or mining areas (Mironenko and Shestakov 1974; Dashko 1987).

In a reverse situation, the increasing aquifer water head causes lowering of the
effective stresses in soils and makes them more susceptible to deformation. This
may cause technogenic (induced) earthquakes in the regions of large-scale water
pumping in deep aquifers or in the construction areas of large water reservoirs
(Osipov 2001). The human-induced increase of the hydrodynamic head in soil and
rock systems is widely applied for hydraulic fracturing of strata and increasing
productivity of oil and gas wells. Recently, the concept of the planned stress
relief in the earth crust using the hydrostastic head has been widely applied in the
development of natural risk reduction methods for seismic regions (Nikolaev 1999).

1.2.2 Hydrodynamic Stresses

Hydrodynamic stress forms in soil when water and other fluids filtrate under the
impact of gravity or other physical fields. *Filtration* usually assumes water movement in a porous medium consisting of pore channels. The kinetic energy of moving water is transmitted to the porous body and changes its stress state.

The stress received by soil from filtrating water depends on the hydrostatic head gradient and the pore space structure. Depending on the water flow direction, an additional stress produced by filtration may either favor the soil compaction or induce shear stresses. Therefore, stress produced by water filtration should be necessarily considered in calculations of slope stability, pressure on retaining walls, etc.

Filtration in a soil mass is most often the result of groundwater level difference, which induces the movement of liquid under gravity. To describe changes in gravitational potential during filtration, let us consider the movement of water represented as a liquid column of l length and ω area (Fig. 1.4). The pressures $P_1 = p_1\,\omega$ and $P_2 = p_2\,\omega$ act at the both column's ends and depend on gravity force $G = \rho_w g\,\omega l$, where ρ_w is water density, g is the gravitational acceleration, and p_1 and p_2 are the pressures at cross-sections 1 and 2, respectively, directed along the column axis.

Introducing function $\varphi = p + \rho_w g z$, where z is the reference point of the flow, and p is the pressure of the flow at point z, we may calculate for cross-sections 1 and 2:

$$\varphi_1 - \varphi_2 = \frac{\theta l}{V}, \tag{1.12}$$

where θ is work of resistance forces, and $\frac{\theta l}{V}$ is the energy required for transportation of a unit of liquid volume between cross-sections 1 and 2.

It follows from Eq. (1.12) that the difference $\varphi_1 - \varphi_2$ represents the specific loss of the water flow energy. Dividing Eq. (1.12) by l, we obtain the gravitational potential gradient:

$$\frac{\varphi_1 - \varphi_2}{l} = I_\varphi = \frac{\theta}{V}. \tag{1.13}$$

In practice, the dependence between the potential φ and the hydrostatic head (H) is used, which describes the potential energy at a given point of the flow and is expressed in linear units:

$$H = \frac{\varphi}{\gamma_w} = \frac{p}{\gamma_w} + z. \tag{1.14}$$

Fig. 1.4 Schematic representation of hydrodynamic head

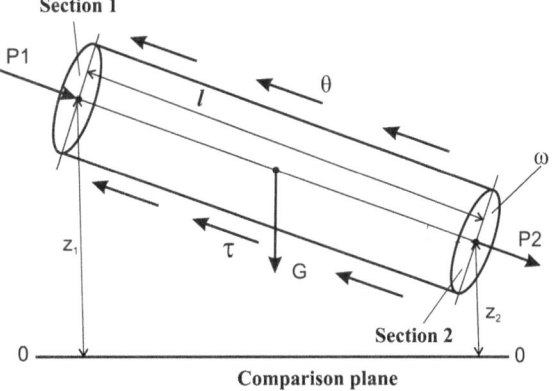

The hydrostatic head depends on height $h_p = \frac{p}{\gamma_w}$ controlling the pressure energy at a point at distance z from the reference plane (Fig. 1.4).

The kinetic energy of flow $E_\kappa = \frac{mv^2}{2}$ is controlled by the velocity head $h_v = \frac{v^2}{2g}$.

D. Bernulli used this equation to prove that the total flow energy may be calculated from *the hydrodynamic head potential* H_d (Mironenko and Shestakov 1974):

$$H_d = H + h_v = \frac{p}{\gamma_w} + z + \frac{v^2}{2g}. \tag{1.15}$$

The flow energy is expended to overcome resistance forces caused by liquid viscosity. The viscous resistance of liquid applied to pore walls is transmitted to the soil skeleton and produces *hydrodynamic stresses* in it (σ_{hd}).

In laminar water flow in a porous body, the flow resistance depends completely on the viscous resistance inside the flow. Isaac Newton has described such flows and stated that the internal friction forces manifested during the displacement of layers are proportional to the relative velocity of their displacement $\frac{dv}{dn}$, where n is the direction normal to the flow. For tangential stress (τ) at viscous friction, the Newton law may be written as follows:

$$\tau = -\mu \frac{dv}{dn}, \tag{1.16}$$

where μ is the dynamic viscosity coefficient.

In the SI system, μ is measured in Pa s (N s/m^2); and in the CGS system, in P (poise). One P is equal to 0.1 Pa s. For water at 20 °C, $\mu = 10^{-3}$ Pa s.

Water filtration in clay has specific features, because clay pores consist mainly of micropores filled largely with bound (adsorbed and osmotic) water and some finer micropores completely sealed by this water. Bound water is affected by the surface forces of particles and has elevated viscosity. Hence, the outermost layers of the bound water films can be involved in viscous flow only upon reaching the initial shear stress (τ_o). The thicker the layer of bound water involved in filtration, the greater the viscous resistance that the moving liquid encounters.

Therefore, in fine pores, the water flow is viscous-plastic; the mutual displacement of water molecules depends not only on their viscous friction, but also on the surface forces of the particles. To describe water movement in such systems, the Bingam-Shvedov's law is used rather than the Newton's law:

$$v = 0 \quad \text{for } \tau \leq \tau_o$$

$$\tau = \tau_o + \mu \frac{\partial v}{\partial n} \quad \text{for } \tau > \tau_o. \tag{1.17}$$

According to Eq. (1.17), the liquid starts moving only after the shear stress τ exceeds the initial shear stress of bound water layers (τ_o).

It appears that the greater the shear stress of the filtrating water, the higher the hydrodynamic pressure transmitted to the skeleton (σ_{hd}). Therefore, the long-term water filtration in poorly lithified clays, in particular, causes their compaction and alignment of the particles perpendicular to the water flow (Berezkina and Tsareva 1968; Pereverzeva and Osipov 1975). Hydrodynamic pressure decreases the ground stability on slopes and induces landslides. Hence, it is important to consider hydrodynamic stresses in the stability calculation of waterlogged slopes. The value and direction of hydrodynamic forces are found analytically, graphically, or from modeling.

Water transfer in porous deposits is controlled not only by gravity but also by other physical fields. From this viewpoint, osmotic filtration is most interesting, as it may be induced by the gradient of salt concentrations and the electric field.

Movement of solvent (water) under the impact of the difference in the concentrations of dissolved salts in the external solution and pores of soil is called *the osmotic filtration* (Mironenko and Shestakov 1974). The osmotic filtration occurs during mass exchange between the internal and external solutions. Water usually moves towards the diffusing salt ions. However, there are exceptions to this rule, as in the so-called normal and abnormal osmosis. In the former case, the osmotic water flow is directed towards the salt concentration gradient; and in the latter case, in the opposite direction, which is related to the number of water molecules in the ionic-hydrate complex of diffusing ions.

Along with the external osmosis, the internal osmosis exists in clays; it is caused by the ion concentration gradient in the diffusive part of the electrical double layer (EDL) of particles and in the pore solution. The difference between the internal and external osmosis is as follows: first, the internal osmosis is manifested only in the local movement of water molecules, towards EDL or in the opposite direction, with the varying thickness of the diffusive EDL part; and second, it does not represent the filtration process. Therefore, we will discuss the internal osmosis in the chapters devoted to internal stress.

During the osmotic filtration, the flow produces the osmotic pressure (P_{osm}), which is transmitted to pore walls. *The osmotic pressure* means the force that causes osmosis per unit surface of an arbitrary semipermeable membrane. The osmotic pressure is usually expressed in MPA or *atmospheres*. The osmotic pressure value for equilibrium systems is found from the Van't Hoff equation:

$$P_{osm} = RTC, \tag{1.18}$$

where R is the ideal gas constant, T is the absolute temperature, and C is the molar concentration of solution.

During the osmotic filtration, similarly to the gravitational filtration, the osmotic pressure is transmitted to the soil skeleton and is aligned with the osmotic filtration. Therefore, depending on the direction of the osmotic filtration, the osmotic pressure may either enhance or mitigate the effective stress in soil.

The osmotic pressure increment velocity is controlled by the gradient of salt concentration in the external medium and in the pore water. It is calculated from the equation:

$$V_{osm} = K_{osm}I_c, \tag{1.19}$$

where K_{osm} is the osmosis coefficient that characterizes the osmotic filtration rate for (unit concentration gradient [cm^5/mol s]) and I_c is the filtration gradient.

Among other physical fields controlling water movement in porous bodies, *electroosmotic filtration (electric osmosis)* is noteworthy. It forms in the presence of the electric field of either natural or technogenic origin in a water-saturated porous medium. The electroosmotic filtration is most typical of clay soils and is caused by transport of hydrated cations from the external layers of the diffusive part of EDL under the influence of the electric field. The velocity of the electroosmotic filtration is proportional to the gradient of electric potential *E:*

$$V_{eo} = K_{eo}E, \tag{1.20}$$

where V_{eo} is the velocity of the electroosmotic filtration, and K_{eo} is its coefficient.

The K_{eo} value depends on the electrokinetic properties of soils, the pore space geometry, and the pore solution properties. Depending on these factors, its value varies between $1 \times 10^{-5} - 8 \times 10^{-5}$ cm^2/V s (Mironenko and Shestakov 1974). The stress in the soil skeleton caused by electroosmosis corresponds to the water mass movement and is similar in origin to other types of filtration.

1.2.3 Stresses Produced by Engineering Structures

Any engineering structure built on the surface or inside a ground body produces stress that is transmitted to ground. Stress distribution with depth, produced by a concentrated vertical load applied to a flat surface is calculated based on the elasticity theory. In this case, it is assumed that the zones of plastic deformations developing at the edges of the loaded area do not exert any noticeable effect on stress distribution with depth. J. Boussinesq used this calculation for the first time ever in 1885 for the vertical concentrated load and elastic semi-space. Several years later, A. Flamant applied this calculation to stress behavior under an evenly distributed load for plane stress conditions.

Stress σ_z induced by the vertical concentrated force (P) applied to the elastic semi-space surface (Fig. 1.5a) is described by the Boussinesq equation:

$$\sigma_z = \frac{K}{z^2}P, \tag{1.21}$$

Fig. 1.5 Schematic representation for calculation of z axis-distribution of stress produced by concentrated (**a**) and distributed (**b**) load

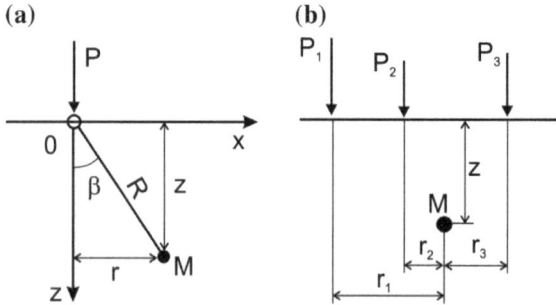

where

$$K = \frac{3}{2\pi} \frac{1}{\left[1 + (r/z)^2\right]^{5/2}},$$

here r is the point coordinate on the xy plane: $r^2 = x^2 + y^2$ and z is the depth of the point relatively to the loading surface.

Using the superposition principle, we may calculate the vertical compressive stress, for example, at point M (Fig. 1.5b), formed by several concentrated forces applied to a surface (Ukhov et al. 1994):

$$\sigma_z = \frac{K_1}{z^2} P_1 + \frac{K_2}{z^2} P_2 + \cdots + \frac{K_n}{z^2} P_n = \frac{1}{z^2} \sum_{i=1}^{n} K_i P_i, \qquad (1.22)$$

where K_i is found from Eq. (1.21) depending on the r_i/z ratio, with z coordinate constant for point M.

The stress distribution with depth z obtained by A. Flamant for the plane problem is:

$$\sigma_z = \frac{2P}{\pi} \frac{z^3}{r^4}, \qquad (1.23)$$

where r and z are the investigated point coordinates.

Calculations conducted with Eq. (1.23) prove that the compressive stresses in points on a semi-plane located along the z-axis decrease hyperbolically:

$$\sigma_z \approx 0.7 P / z. \qquad (1.24)$$

Field measurements of stress in cohesive soil masses prove that they fit, to a certain degree, the results of the calculations based on Eqs. (1.23) and (1.24).

The depth-dependent stress values calculated for soil masses may be presented as equal-stress lines (stress isolines). Figure 1.6a shows the left part of the symmetrical pattern of the vertical stress σ_z isolines. The stress intensity decays gradually with the growing distance from the loaded surface. The vertical stress isolines (σ_z) are spread mainly down the mass's thickness, while the horizontal stress isolines σ_x run sideways from the loaded area (Fig. 1.6b) and the tangential stress isolines (τ_{xy}) are concentrated at the edges of the loaded area (Fig. 1.6c).

The distribution pattern of the vertical stresses in soil depends on the area and shape of the foundation transmitting stress to the soil. For example, Fig. 1.7 shows the stress distribution with depth for the loaded areas of various shapes: (1) square ($l = b$), (2) band ($l \geq 10b$) of b width, and (3) band of $2b$ width. It is clear that stress decreases with depth quicker under the loaded square area (curve 1) than under the flat elongated area (curve 2). Increase of the width of the band in a two-dimensional problem leads to an even slower decrease of the stresses with depth (curve 3) (Ukhov et al. 1994).

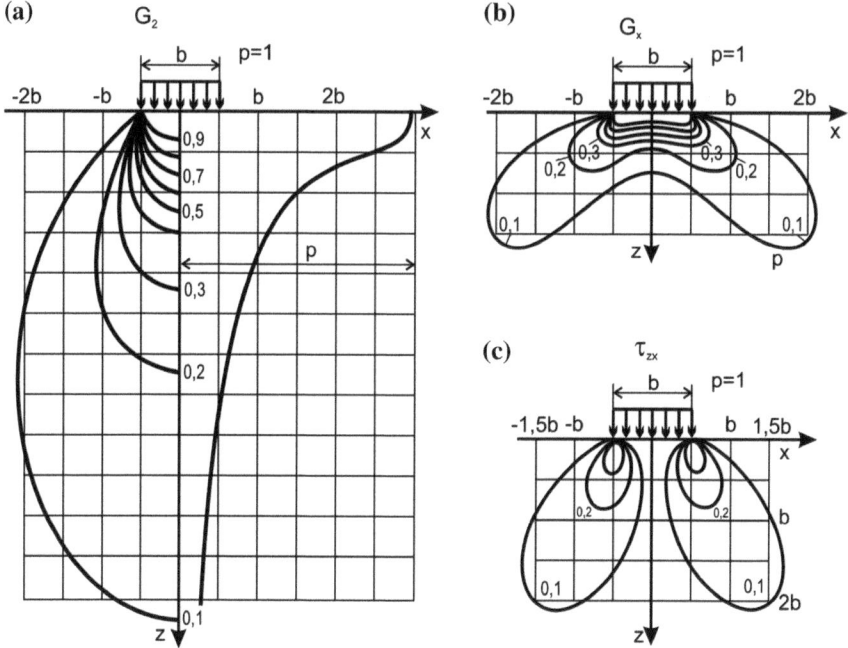

Fig. 1.6 Stress distribution in soil produced by uniform load: **a** isolines and distribution diagram of vertical compression stress σ_z, **b** isolines of horizontal σ_x and **c** tangential σ_{xz} stress (Ukhov et al. 1994)

Fig. 1.7 Stress σ_z distribution as a function of shape and area of a loaded plate: *1* square; *2* strip; *3* wide strip (Ukhov et al. 1994)

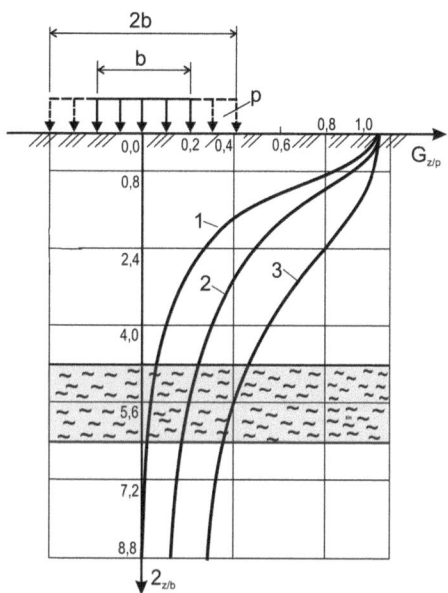

The discussed laws have practical significance. For instance, they are used when a weak layer is present at a certain depth beneath an engineering structure (Fig. 1.7). The calculations allow selecting the area and shape of a foundation to minimize the stress from an engineering structure on the roof of a weak layer.

1.3 Internal Stress

In addition to external stresses, there are internal stresses in fine-grained soil systems. They are induced by physicochemical processes at any local point in a soil mass; they result from various surface phenomena at the mineral–air and mineral–liquid boundary. These phenomena include the formation of EDL and the electrostatic field of particles, the local osmotic process, the formation of bound-water films on particles' surface, the formation of capillary menisci, recharge of clay particles' sheared edges at different pH, etc. Stresses caused by the above-mentioned processes are discussed below.

1.3.1 Capillary Pressure

The *wetting* phenomenon has the atomic-molecular origin and is observed in liquid-solid interactions. It implies that angle θ, i.e., the *wetting angle*, is formed between the liquid surface and the solid body.

Angle θ is the measure of wetting. The more a drop spreads on the surface, the lower is the θ value, and, hence, the higher is wetting. When a liquid spreads on a solid surface to form $\theta < 90°$, the surface is called *hydrophilic*, and for $\theta > 90°$, *hydrophobic*.

Wetting is the reason for existence of capillary menisci in pores and capillary moisture condensation. *Capillary condensation* occurs when a liquid surface warps and a meniscus is formed at the contact with gas or other liquid. Condensation appears because the relative vapor tension above the concave meniscus is lower than that above the flat surface. According to the *Thomson law*, the pressure of saturated vapor (P_0) above a concave meniscus is a result of changes in the surface tension of liquid:

$$P_0 = P_s \exp\left(-\frac{2\sigma_w \sigma V_{mol}}{R_0 RT}\right), \tag{1.25}$$

where P_s is the pressure of saturated vapor over the flat surface, σ_w is the surface tension of liquid, V_{mol} is the molar volume of liquid, R is the universal gas constant, T is the absolute temperature (K), and R_0 is the capillary radius.

The pressure difference over the warped (P_o) and flat (P_s) surfaces produces *capillary pressure* (P_c) described by *the Laplace formula:*

$$P_c = P_o - P_s = \frac{2\sigma_{(w)}}{r} = \frac{2\sigma_{(w)} \cos \theta}{R_0}, \qquad (1.26)$$

where r is the curvature radius of the meniscus; σ_w is the surface tension of the liquid, θ is the wetting angle, and R_0 is the capillary radius.

The capillary pressure is considered positive for a concave meniscus and negative for a convex meniscus.

Capillary menisci and capillary pressure are formed during vapor condensation in soil pores and their relative saturation $P/Ps \approx 1$. This is often observed in soils in the aeration zone. The formed menisci draw particles close together and increase the structural strength of soil.

A meniscus of r_o radius at the contact of two particles represents a rotation surface characterized in each point by the curvature radii r_1 and r_2, with $1/r_1 + 1/r_2 = const$ (Fig. 1.8a). On a hydrophilic surface, the attraction (adhesion) forces between water molecules and the solid surface (F_a) act at any point of the meniscus:

$$F_a = -\pi r_1^2 \sigma_w \left(\frac{1}{r_1} + \frac{1}{r_2} \right), \qquad (1.27)$$

together with the cohesive forces F_c produced by the liquid surface tension acting within the entire wetting perimeter:

$$F_c = 2\pi r_1 \sigma_w, \qquad (1.28)$$

All these forces produce the resultant force F that binds the solid surfaces thus providing for *the capillary contracting forces* between particles (Fig. 1.8):

$$F = F_a + F_c = \pi r_1 \sigma_w \left(1 - \frac{r_1}{r_2} \right), \qquad (1.29)$$

The contracting force F depends on the amount of water in a pore. With a decreasing amount of liquid in pores during clay drying, the capillary contracting force increases to reach the maximum at $r_1 \to 0$. The capillary contracting force

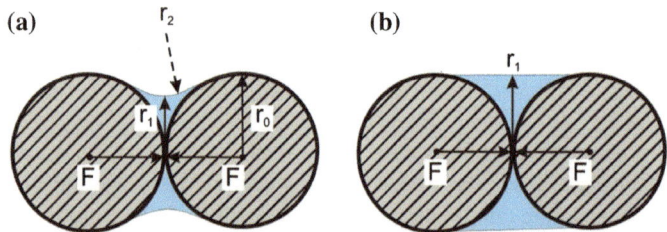

Fig. 1.8 The contracting effect of a water meniscus between two spherical particles

decreases with increase of the amount of liquid and disappears completely when pores are filled with water (Fig. 1.8b).

Further moistening results in the filling of capillary pores with water and capillary uplift of water along the capillary pore network that forms vertical channels (Fig. 1.9a). The menisci of pore angles disappear and are replaced by the menisci of capillary pores (Fig. 1.9b) that control the lifting force and the height of capillary uplift. The lifting force of menisci is transmitted to the soil skeleton and increases effective stress.

It is clear that in an isotropic medium, the developing capillary pressure is equal in all directions. An experiment with a ball-shaped body formed from dry clay powder by its point wetting in the mass center illustrates this statement. To assess quantitatively the capillary forces acting in the capillary zone, it is necessary to calculate their effect in the direction normal to one of the arbitrary planes. This calculation uses the capillary uplift value and the mass of water capillary uplifted above its free surface.

The height of the capillary rise (h_c) may be calculated from the Laplace formula:

$$h_c = \frac{2\sigma_w \cos\theta}{R_0 \rho_w g},\qquad(1.30)$$

where R_0 is the capillary radius equal to $R_0 = r\cos\theta$; r is the meniscus curvature radius; σ_w and ρ_w are the surface tension and liquid density, respectively; and g is the gravity acceleration. For calculations, the R_0 value is taken equal to the average pore dimension.

Since the entire raised water mass transmits its weight to pore walls, it determines the total stress produced by capillary forces in the soil skeleton:

$$\sigma_c = m_w h_c g,\qquad(1.31)$$

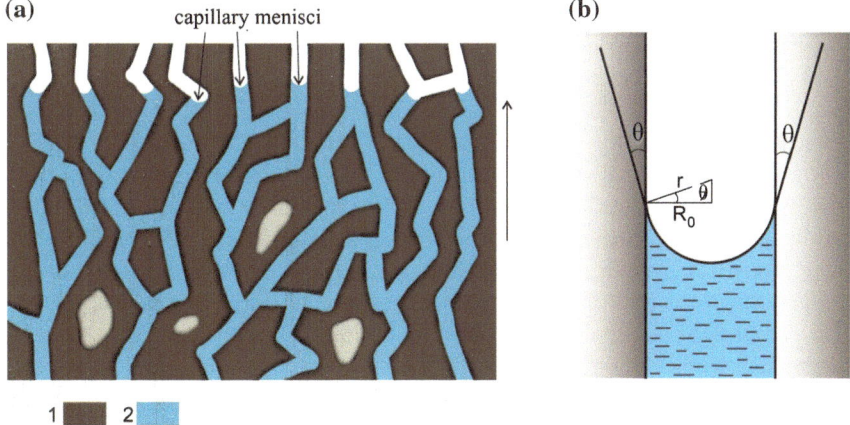

Fig. 1.9 Water uplift by capillaries (**a**) and a water meniscus in a tube pore (**b**): *1* pore walls, *2* capillary water

where m_w is the mass of capillary uplifted water (thereafter, capillary water) in the soil unit volume, h_c is the capillary uplift height, and g is the gravity acceleration.

In sand, the height of capillary uplift ranges from several cm to 1 m. Calculations based on Eq. (1.31) show that the maximal total stress, which may develop due to capillary forces in sand, reaches some kPa/cm^2. In clay, σ_c is higher.

Thus, the value and manifestation character of capillary forces depend on the water amount in soil. The contracting force of pore angle menisci is formed in the contact zones between particles when moisture content reaches the maximal hygroscopic moisture (w_{mg}). Further wetting of soil results in filling capillary pores with osmotic and capillary water and induces capillary uplift along tubular capillary pores. With the capillary rise, the contracting stress caused by the weight of capillary water increases and replaces the forces of pore angle menisci. This effect increases until moisture reaches capillary water capacity (w_c), after which the capillary forces start decaying and disappear completely when the total water saturation of soil (the moisture of total water capacity—w_{sat}) is reached.

The discussion presented above suggests that if the soil occurs in the aeration zone above the capillary fringe, the stress produced by the capillary forces should be calculated from Eq. (1.29). If the soil is within the zone of capillary moistening, i.e., above the ground water table but below the capillary fringe, the capillary stress is calculated according to Eq. (1.31). No capillary forces exist in soils below the ground water table.

1.3.2 Disjoining Pressure of Hydrate Films: Components of Disjoining Pressure

Dispersed systems exist in nature due to the presence of thin interphase hydrate films on the surface of mineral particles. Hydrate films control the specific properties of fine-grained systems, i.e., the capacity to aggregation or disintegration, plasticity, swelling capacity, compressibility, and many others.

According to kinetics principles, stability or instability of microheterogenic disperse systems is controlled by the balance between the forces of attraction and repulsion between particles. Attraction arises from the Van der Waals forces of intramolecular interaction. Repulsion forces result from the electrostatic interaction when the outer (diffusive) parts of EDLs of particles overlap. The formation of the adsorbed water film on the mineral surface of particles with specific structural properties is another reason preventing particle from being drawn closer.

The thickness of stable boundary water films in the contact gap depends on the thermodynamic equilibrium between the attraction and repulsion forces in interacting bodies. Any disturbance of the thermodynamic and, concurrently, of the mechanical equilibrium between particles causes changes of the hydrate layer thickness, i.e., requires applying additional force to the film surface. This additional force counterbalances the pressure in the thin film and maintains equilibrium in the system.

The excess pressure that appears in the interphase liquid layer during its thinning or thickening is called disjoining pressure ($\Pi(h)$).

The notion of disjoining pressure in thin boundary films was first introduced by Russian scientists Deryaguin and Kusakov (1936). B.V. Deryaguin continued to develop this theory in collaboration with N.V. Churaev, S.V. Nerpin, E.V. Obukhov, Z.M. Zorin, A.S. Titievskaya, G.A. Martynov, etc.; international scientists contributed as well (e.g., J. Israelachvili, S. Marčelja, N. Radic, R. Pashley, J. Kitchener, etc.).

Thermodynamically, disjoining pressure of hydrate films is the difference between the pressure in the film between solid surfaces and the pressure in the water that is in the thermodynamic equilibrium with the film (Deryaguin 1955). $\Pi(h)$ is equal to the difference between the pressure on hydrate film exerted by the bounding solid phases (p_1) and the pressure of the liquid phase (p_0) outside the film:

$$\Pi(h) = p_1 - p_0 \tag{1.32}$$

Disjoining pressure depends on the degree of overlapping of hydrate films, i.e., on film thickness h in the contact gaps. In turn, h is the function of the mutual action of attraction and repulsion between solid surfaces separated by a liquid interlayer.

Function $\Pi(h)$ depends on the cumulative effect of the attraction and repulsion forces between the particles; *the free surface energy of bilateral film* (ΔU) is correlated to the cumulative effect. The ΔU value per surface unit area is equal to the doubled value of the specific interphase energy σ at the interphase boundary: $\Delta U = 2\sigma$. The relationship between disjoining pressure ($\Pi(h)$), the film energy, and film thickness is as follows:

$$\Pi(h) = \frac{d\Delta u(h)}{dh}. \tag{1.33}$$

The ΔU value is measured in J/m^2 and represents the excess free energy per film unit area.

Disjoining forces may be evaluated for surfaces of various shapes. Deryaguin et al. (1985) derived the equation for calculation of interaction forces of any origin for curved surfaces of arbitrary shape:

$$F(h) = \pi k U(h) \tag{1.34}$$

where k is the geometric parameter controlled by the curvature of contacting surfaces. For example, for two spherical particles with the radii r' and r'', $k = 2r'r''(r' + r'')$.

Disjoining pressure may be either positive, i.e., preventing the film thinning, or negative, i.e., promoting the film thinning. It is important to note that disjoining pressure creates normal stress always directed perpendicular to the surfaces, between which it acts. In addition, it does not relate to either viscosity or to other mechanical properties of liquid. Changes in film thickness develop so slowly, that energy dissipation and work expended in overcoming the liquid viscosity may be neglected.

The value of disjoining pressure $\Pi(h)$ and the thickness of boundary films h are controlled by the input of surface attraction and repulsion forces of various origin. To a first approximation, this input may be considered additive, consisting of the following components (Deryaguin and Churaev 1984):

$$\Pi(h) = \Pi e + \Pi m + \Pi s, \qquad (1.35)$$

where Π_e is the electrostatic component resulting from overlapping of diffusive ionic layers of charged surfaces of particles and their repulsion, Π_m is the molecular component produced by the dispersion interaction of the solid surface through a thin liquid film, and Π_s is the structural component of disjoining pressure caused by overlapping of adsorption liquid layers whose structure is transformed.

Assessment of each component is based on the theoretical achievements in molecular physics. The stability of films is controlled by the manifestation patterns of each disjoining pressure component. The contribution of the individual components can be found from the equations describing the force of specific interactions per unit area.

1.3.2.1 Molecular Component

Surface forces of molecular origin act at the solid-liquid-gas interphase boundary. The total value of these forces depends on the total area of the interphase surface; it grows with the increasing dispersion of the system. Therefore, molecular interaction plays a very important part in clay formations. Molecular (or Van der Waals) forces are essential in coagulation and sedimentation of clay particles, as well as in cohesion formation in mud and poorly lithified clay.

The general theory of molecular interactions was originally formulated by F. London in 1937. Later, H.C. Hamaker, W. Verwey, and G. Overbeek contributed significantly to its development by calculating molecular attraction between fine-grained particles.

For two particles separated by a thin flat gap (h) filled with water, the attraction energy per the gap unit area (U_m) and the value of the molecular component of disjoining pressure (Π_m) can be found from the following equations:

$$U_m = -\frac{A^*}{12\pi h^2} \qquad (1.36)$$

$$\Pi_m = -\frac{A^*}{6\pi h^3}, \qquad (1.37)$$

where A^* is the complex Hamaker constant that depends on the polarization capacity of molecules, as well as on the density of film and condensed bodies that it separates. The complex Hamaker constant decreases as the nature of interacting phases (the hydrate film and the fine-grained medium) becomes closer. If the chemical composition and the structure of interacting phases are close, the A^* value may reach 10^{-21} J.

Thus, for symmetrical films, the molecular component of disjoining pressure is negative. This means that the molecular forces tend to make the hydrate film separating the solid surfaces thinner.

The energy and force of molecular interaction between fine particles should be assessed in relation to pair-wise interaction of two particles rather than to the interlayer unit area. Hence, the force of molecular interaction of particles is controlled not only by the distance between the particles and the Hamaker constant, but also by their size and shape. Thus, the calculations based on the microscopic theory developed by F. London prove that the U_m value, for instance, is inversely proportional to h^2 for two thick plates ($\delta \gg h$), where δ is the plate thickness, and it is inversely proportional to h^3 for thin plates ($\delta \gg h$).

Calculations show that for the particles with radius $r \approx 10^{-8}$ m (10 nm), film thickness $h \approx 2 \times 10^{-10}$ m, and the Hamaker constant $A^* = 10^{-19}$, the energy of molecular interaction is 4×10^{-19} J, which is approximately 100 times higher than the energy of thermal motion of particles (kT).

Later, mainly in the works by E.M. Lifshitz, I.E. Dzyaloshinskii, and L.P. Pitaevskii, the theory of molecular forces received different interpretation and was named *the macroscopic theory*. According to the macroscopic theory, the calculations based on the F. London theory and assuming the additivity of the pair-wise molecular interactions are applicable only for short distances between the bodies. With the growing distance, these calculations become invalid, as they do not take into account the so-called *retardation effect of molecular forces*. If the gap between the bodies exceeds λ_0, i.e., the wave-length that corresponds to the transition between the basic and excited states of an atom, the F. London theory becomes invalid due to the existing terminal velocity of electromagnetic wave propagation. As a result, at large distances, the dispersion forces decay faster than is predicted by the F. London theory. The energy of molecular attraction becomes inversely proportional to h^2, h^3, and h^4 for spherical particles, thick plates, and thin plates, respectively.

The presence of molecular forces between solid bodies was established experimentally in a sophisticated physical experiment. This study was pioneered by Deryaguin and Abrikosova (1951). They used a quartz plate, of area 4×7 mm, and quartz spherical lenses with the curvature radii R_0 ranging from 5 to 26 cm, whose surfaces were thoroughly polished and purified from pollutants and dust. The distance between the plate and a lens was measured with accuracy of up to ± 0.01 μm. Molecular attraction was measured by an electric compensation device rigidly bound to the beam of a precision balance providing the weighing accuracy of up to 10^{-5}–10^{-6} g. The attraction force was calculated from the current necessary to keep the interacting bodies at a permanent distance h:

$$F(h) = -2\pi R_0 U(h) \tag{38}$$

The measurements allowed the researchers to obtain the common relationship between U and h for the plate and three lenses with the radii $R_0 = 11.1$, 10, and 25.4 cm (Fig. 1.10). The first measurements of molecular attraction energy for the minimal distance between the bodies (0.1 μm) produced the result of 3×10^5 erg/cm^2 (40 mJ/m^2) (Deryaguin and Abrikosova 1951).

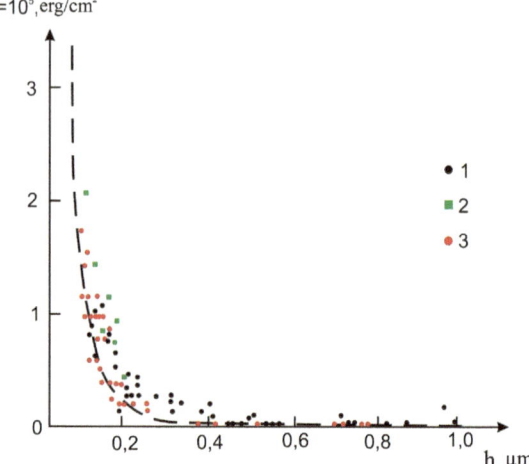

Fig. 1.10 Dependence between molecular attraction energy and distance (per unit area) for a quartz plate and lenses of different radii R_0: (*1*) 11.1, (*2*) 10, and (*3*) 26 cm (Deryaguin and Abrikosova 1951)

Later, similar measurements were performed by Rabinovich (1977), Shchukin et al. (1969), Babak et al. (1977, 1984), and an Australian scientist Israelachvili (1978). The latter has developed a device using atomically smooth thin mica plates curved for the measuring of interaction between cylinders positioned cross-wise. The energy of molecular interaction between two silicate plates was estimated to be 0.01–10 mJ/m^2.

1.3.2.2 Electrostatic Component

Particles of clay minerals have electrical charges compensated by the counterions located on the internal and external faces of crystallites. Thus, EDL is formed on the surface of particles whose internal part consists of the charged mineral surface and the external part consists of the compensating counterions. In the absence of moisture in the environment, the compensating ions neutralize the particle charge completely, which provides its electrical neutrality.

In damp air, or during the direct contact between a particle and a liquid, EDL undergoes change, i.e., compensating ions are hydrated, which weakens the interaction with the particle surface. For some ions, the energy of interaction with the mineral surface becomes lower than the energy of thermal motion ($ze\varphi_0/4kT$), which results in "deterioration" of the ionic cover of the particles and partial diffusion of ions over some distance from the mineral surface. The remote ionic complexes do not lose ties with the particle but remain within its electrostatic field. Thus, in an aquatic medium, some counterions remain near the mineral surface and form *the adsorption layer of counterions,* i.e., the Schtern-Helmholtz layer, while other ions move away from the surface and form the external *diffusive layer of counterions*, or the layer of Gui-Chapman. Since clay particles are negatively charged, both adsorption and diffusive layers of EDL retain cations, while anions are removed to the free solution.

Concurrently with the formation of the diffusive part of ions in EDL, the amount of water molecules attracted to the diffusive layer by osmotic forces increases. The osmotic moisture transfer is caused by the difference in the cation concentrations in EDL and in the pore solution, which separates this process from the osmotic filtration. EDL acts as a semipermeable membrane. The osmotically attracted molecules are involved in the hydrate complex of the counterions and form a second hydrate layer of bound water.

The diffusive ion layer forms an electrostatic field around the particles spreading to some distance from the particle surface. This field's potential (ψ_δ, the Schtern potential) decreases gradually with the growing distance from the particle surface compensated by the ions of the diffusive layer. Outside the electrostatic field of the particle, the external solution (electrolyte) is symmetrical, i.e., the number of cations is equal to that of anions. The diffusive layer is much thicker than the adsorption layer and is distributed from the particle surface over a distance of more than 100 nm.

If two fine-grained particles, separated by a gap of h width, get closer to each other, their ionic atmospheres overlap. The potential is redistributed in the gap between clay particles, becoming higher than the potential of individual particles. In the gap center, the potential ψ_M is approximately double the value of a single particle's potential at the same $h/2$ distance from the surface $\psi_{0(h/2)}$ (Fig. 1.11).

$$\psi_m = 2\psi_{0(h/2)}. \tag{1.39}$$

The rising concentration of ions in the intergrain gap in the overlapping ion atmospheres creates a local osmotic pressure that forces the liquid to flow between the particles and disjoin them. As a result, the so-called *electrostatic component of disjoining pressure* (Πe) appears.

The theory of electrostatic interaction of diffusive ionic layers surrounding fine-grained particles was developed by B. Deryaguin, L. Landau, W. Verwey, G. Overbeek and many others in the 1930–1940s. At present, it is the foundation of the quantitative theory of stability of highly-dispersed systems (DLVO theory).

Fig. 1.11 Distribution of potential in a gap between two parallel and similarly charged plane surfaces at a constant surface charge ψ_0

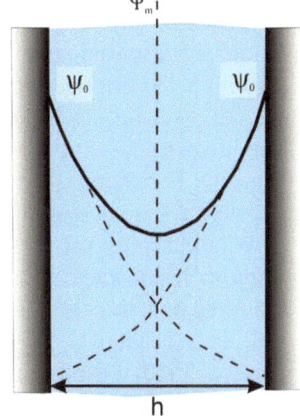

According to this theory, the electrostatic disjoining energy increases with thinning of the dividing gaps between particles. The repulsion energy of two plane-parallel particles with low surface potential (ψ_0) is found from the following equation (Shchukin et al. 2007):

$$\Pi_e \approx \frac{4z^2 e^2 \psi_0^2 n_0}{kT} e^{-\text{æ}h},\tag{1.40}$$

where z is the charge of counterions, e is the electron charge, kT is the energy of heat movement of ions, æ is the parameter characterizing the ion atmosphere thickness $\delta = 1/\text{æ}$, and h is the thickness of the gap between particles and n_0 is the concentration of ions in the gap.

According to Eq. (1.40), disjoining pressure Π_e is proportional to the surface potential ψ_0 squared.

For a highly charged surface, with the surface potential $\psi_0 > \frac{4kT}{ze}$, the electrostatic component of disjoining pressure does not depend on ψ_0 value and can be found from the following equation (Shchukin et al. 2007):

$$\Pi_e \approx 64 n_0 kTe^{-\text{æ}h}.\tag{1.41}$$

Integration of the electrostatic component of disjoining pressure over the gap thickness allows us to estimate its energy variation (Shchukin et al. 2007):

$$U_{(e)} \approx \frac{64 n_0 kT\gamma^2 e^2}{ae} e^{-\text{æ}h},\tag{1.42}$$

where $\gamma \approx \frac{ze\psi_0}{4kT}$.

Unlike the molecular forces, electrostatic forces are positive and cause particle repulsion.

A more strict analysis of the electrostastic repulsion energy variation between the charged planes with the decreasing gap between them gives the following dependence (Shchukin et al. 2007):

$$\Pi_e \approx 2 n_0 kTch\left[\frac{ze\psi_m(h/2)}{kT}\right] - 2 n_0 kT\tag{1.43}$$

The physical meaning of this equation implies the following: the first summand stands for the osmotic pressure inside the gap and the second one represents the osmotic pressure in the dispersion medium caused by the external (filtration) osmotic pressure. In practice, it is virtually impossible to separate these two constituents. Therefore, osmometers applied for this purpose show the integral effect of external and internal osmosis.

With changing pH of a medium, the electrostatic interaction acquires specific features. In an acidic medium, a particle's edges become positively charged and the faces remain negatively charged; the total potential of the particles decreases and, as a result, Π_e decreases too; face-to-edge contacts appear (Shofield and Samson 1953; Osipov and Sergeev 1972; Osipov 1973; Osipov and Sokolov 2013) (Fig. 1.12). On the contrary, in an alkaline medium, the surface potential of

Fig. 1.12 Schematic representation of interaction between two particles in an acidic medium

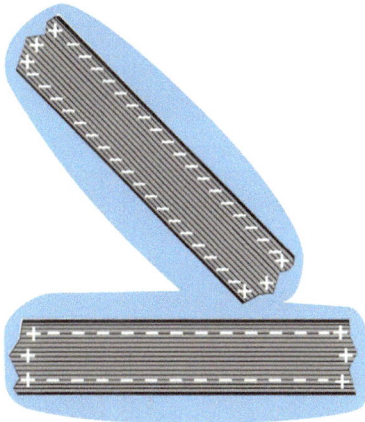

particles increases due to the negatively charged faces thus promoting increase of Π_e and stabilization of the particles.

1.3.2.3 Structural-Mechanical Component

Clay minerals interact readily with water. Even in the air, they adsorb water molecules and retain them physically bound. This phenomenon is based on the adsorption of H_2O molecules on mineral surfaces. A thin water film, with specific structure and properties, formed under the impact of forces on a mineral surface is referred to as *the adsorption film*.

The mechanism of water molecule adsorption is based on the epitaxy phenomenon, i.e., adaptation of water molecules to the geometry of surface adsorption centers, deformation of the hydrogen bonds, and formation of the hydrate cover with a transformed structure (compared to free water) (Ravina and Low 1972; Osipov 1976, 2011, 2014; Pashley and Israelachvili 1984). The influence of the active surface centers weakens with thickening film of the adsorbed molecules, which gradually decreases the deformation of hydrogen bonds and provides for the water transition from the deformed to regular structure.

The adsorbed water molecules are retained firmly on the mineral surface. They are considered "pressed" to a solid body by surface forces and to be in a specific stress state. The anisotropic stress in these films produced by the deformation of hydrogen bonds controls their specific properties, i.e., they do not transmit the porous pressure and do not manifest the hydraulic uplift. In addition, the adsorption films have specific structural-mechanical properties, i.e., the elevated viscosity and disjoining pressure (Churaev et al. 1971; Zorin et al. 1972; Deryaguin et al. 1972; Churaev and Yashenko 1973; Deryaguin and Churaev 1984). The water viscosity in a 3–4 nm thick adsorption film is 0.003–0.005 Pa s (at 23–30 °C), which is 3–5 times higher than the free water viscosity. It also differs from free water in its dissolving capacity, thermophysical, dielectric, and other properties.

Fig. 1.13 Generation of repulsion forces between two flat surfaces in overlapping adsorbed water films: *1* particle, *2* osmotic water, *3* adsorbed water, *4* overlapping zone of adsorbed water films

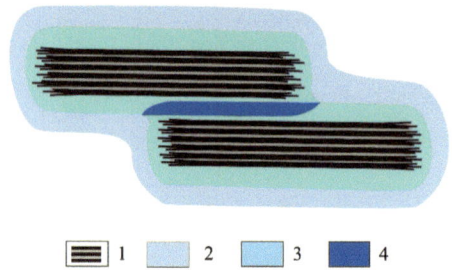

When clay particles get closer together and adsorbed water films overlap, the repulsion forces caused by the structural-mechanical properties of the adsorbed films emerge in the contact zones of the particles (Fig. 1.13). The destruction of the adsorbed water structure in the overlap zone is accompanied by changes in free energy and developing disjoining pressure. The developing forces that prevent contact of particles were termed *the structural component of disjoining pressure* Π_s (Deryaguin and Churaev 1984; Deryaguin et al. 1985; Deryaguin 1986).

The existence of the structural disjoining pressure and its dependence on the thickness of the adsorbed film were proved experimentally by Deryaguin and Zorin (1955) and Deryaguin and Churaev (1972) for wet surfaces of quartz and glass. Later, similar data were obtained for mica plates (Israelachvili and Adams 1978) and montmorillonite (Viani et al. 1983). Australian physicochemists Israelachvili and Pashley (1983) relate the structural disjoining effect to cation-hydrating water molecules and term it *the hydration effect*.

The action radius of structural forces depends considerably on the hydrophilic properties of substrate and the properties of liquid. For hydrophilic surfaces, it may reach some tens of nm (Pashley and Kitchener 1979). We may consider that the structural component of disjoining pressure in clay systems acts in thin films of $h < 5$ nm thick. Hydrophobization of surface shortens noticeably the action radius of structural forces up to their complete disappearance.

Marčelya and Radič (1976) were the first to calculate disjoining pressure of adsorption films Π_s using the empirical relationship that they have deduced:

$$\Pi_s = K/sh(h/(2l)),\qquad (1.44)$$

where l and K are empirical parameters.

For $h > l$, the structural repulsion forces are exponentially dependent on the gap thickness:

$$\Pi_s = 4K\exp(-h/l)\qquad (1.45)$$

K and l values were tabulated by Deryaguin and Churaev (1984) through analysis of experimental data obtained by various authors for glass, quartz, and mica surfaces (Table 1.1). Despite the simplifications assumed, this Eq. (1.45) gives the dependence between Π_s and the gap thickness close to the experimental results.

Table 1.1 Values of K and l constants for water

Mineral	K (dyne/cm²)	l (nm)	Authors
Glass and quartz	9.94×10^7	2.33	Deryaguin and Zorin (1955)
Quartz	7.3×10^6	1.01	Pashley and Israelachvili (1984)
Mica	10^8	1.0	Israelachvili and Adams (1978)
Montmorillonite	2×10^7	2.2	Viani et al. (1983)

1.3.3 Isotherms of Disjoining Pressure of Hydrate Films

The above-discussed molecular, electrostatic, and structural components of dis-joining pressure act concurrently in the hydrate film separating particles and fol-low different laws depending on film thickness h (Fig. 1.14). The molecular component has a negative value, i.e., it tends to draw particles together, and this is manifested to some extent at any thickness of boundary film. Electrostatic and structural components have positive values and prevent particles from being drawn closer.

The electrostatic component (Π_e) has the maximal range of action, reaching 300 nm in dilute solutions. Therefore, thick water films are stable in dilute elec-trolytes because the electrostatic component Π_e is mainly controlled by the surface charge and depends on temperature to a lesser extent.

The molecular component (Π_m) is most effective for film thickness <50 nm. The range of action of molecular forces increases with the growing difference in refraction indices of the liquid and substratum.

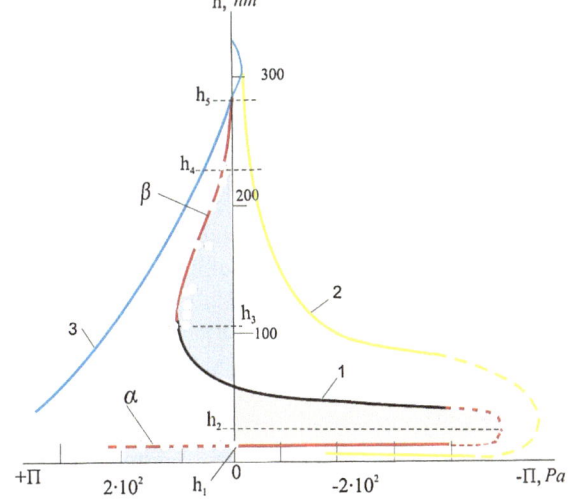

Fig. 1.14 The integral isotherm of attraction and repulsion forces at contacts between clay particles: *1* integral isotherms, *2* molecular forces of attraction, *3* electrostatic forces of repulsion, α and β disjoining pressure of α- and β-films; h_1–h_5 distance between particles; *I* and *II* far- and near-field coagulation contacts; • experimental points (Deryaguin et al. 1985)

For film thickness <10 nm, the disjoining effect emerges due to the structural component of disjoining pressure (Π_s).

Isolation of disjoining pressure produced by the structural component Πs in the total disjoining effect is associated with certain difficulties due to an inadequate level of theoretical knowledge in this area. Hence, experimental data are used for quantitative estimates of the structural component.

The overall dependence between the excess free energy of the boundary film in the interparticle gap caused by the impact of the molecular and electrostatic components is calculated by summing Eqs. (1.42) and (1.36):

$$\Delta U(h) = \frac{64 n_0 kT \gamma^2 e^2}{\ae} e^{-\ae h} - \frac{A^*}{12 \pi h^2} \qquad (1.46)$$

A similar expression for disjoining pressure may be obtained by summing Eqs. (1.40), (1.41), and (1.37). For the particles with a low surface potential (ψ_0):

$$\Pi(h) = \frac{4 z^2 e^2 \psi_0^2 n_0}{kT} e e^{-\alpha e h} - \frac{A^*}{6 \pi h^3}, \qquad (1.47)$$

and for the particles with a high surface potential:

$$\Pi(h) = 64 n_0 kT e^{-\ae h} - \frac{A^*}{6 \pi h^3}, \qquad (1.48)$$

where z is the charge of counterions, e is the electron charge, kT is the energy of thermal motion of ions, $\ae$ is the parameter characterizing the thickness of the EDL ion atmosphere ($\delta = 1/\ae$), h is the gap size between particles, n_0 is the concentration of ions in the gap, and A^* is the complex Hamaker constant.

The dependence between the total action of disjoining forces in the boundary hydrate film in the gap between solids and film thickness h is called *the disjoining pressure isotherm* (Deryaguin and Churaev 1984). Figure 1.14 (line 1) shows the qualitative pattern of this relationship.

For the isotherm plotted in Fig. 1.14, stable films have thickness ranging from h_1 to h_2 and h_3–h_5. The film stability intervals h_3–h_5 and h_1–h_2 are called β- *and* α-*branches of the disjoining pressure isotherm* (Deryaguin and Churaev 1984). The first branch corresponds to thick films with h reaching hundreds nm, while the second one corresponds to thin films with thickness rarely exceeding 10 nm. The isotherm interval from h_1 to h_3 corresponds to the unstable state of films.

The β-branch (β-film) depends mainly on the electrostatic component of disjoining pressure Π_e. The state of β-film is controlled by the structure of the diffusive part of EDL; this state is highly sensitive to physicochemical conditions. The thickness of β-film grows with the decreasing concentration of ions in the outer solution as well as with the decreasing valence of ions in the diffusive layer; on the contrary, β-film may become unstable and disappear in highly concentrated solutions. Unlike α-film, the disjoining action of β-film is almost independent of temperature, which indicates the different origin of these films (Deryaguin and Churaev 1972, 1984; Deryaguin 1986).

The α-branch of the isotherm (α-film) is more thermodynamically stable. It is formed by the adsorbed water layer in equilibrium with the saturated vapor of volume liquid at $P/P_s \approx 1$, i.e., corresponds to the maximal number of water molecules in the adsorbed state. Thermodynamic stability of α-film is controlled by the crystallochemical properties of the solid surface as well as by the composition of ions in the EDL adsorption layer. Increasing density and energy of surface adsorption centers and the geometry of their location on faces of clay minerals influence the epitaxy development and, hence, the thickness and stability of α-film. A growing concentration of ions in the adsorption layer and their valence destruct α-film and decrease its thickness and stability (Aronson and Princen 1978; Deryaguin and Churaev 1984).

The temperature dependence of α-film proves the structural nature of disjoining forces: the α-film thickness decreases quickly with temperature rise and disappears upon reaching $t \approx 65$ °C (Perevertaev and Metsik 1966; Deryaguin 1986).

The metastable β-film may break under the external pressure or physicochemical factors. The applied pressure or a high concentration of cations in the external solution compensates for the film disjoining pressure and thins the film. Under the impact of these factors, disjoining pressure of β-film reaches its minimal critical thickness h_3 and the film raptures. The further disjoining effect at the contact between two bodies is exerted by α-film, in which the structural component plays the main role.

A question on how stresses transmitted to hydrate films are transformed is intriguing. It is known that bound water, similarly to free water, does not manifest the shear strength. At the same time, it shows elevated viscosity. Thus, bound water is slowly squeezed out of the contact zone. Since EDL of particles exists universally in the compensated state, the squeezed-out water transitions from the diffusive layer to the free state. This increases the pore pressure.

A similar effect is also observed when the bound water α-film with specific structure raptures at the contact raised areas. However, unlike the osmotic β-film water, the adsorption α-film water is not squeezed out, but loses its specific structure and becomes regular free water. In a water-saturated system, this also leads to an increase of the pore pressure.

It is important to note that bound water of α- and β-films does not transmit hydrostatic and pore pressure, because it is under the impact of more powerful compressive surface and structural forces. Therefore, it appears virtually impossible to measure the pore pressure in closed pores of clay sealed with bound water films by the routine methods of soil mechanics.

1.3.4 Experimental Studies of Disjoining Pressure of Hydrate Films

Theoretically, the disjoining pressure isotherm $\Pi(h)$ may be calculated only for some cases and with limited accuracy. To obtain more reliable data, the experimental methods are applied for determining disjoining pressure of α- and β-films.

The first experiments on finding disjoining pressure were performed by Deryaguin and Kusakov (1936). They measured pressure in a thin hydrate film at the contact between a solid surface and an air bubble by creating specific conditions under which floating air bubbles of various radii R_0 approached a smooth horizontal surface (Fig. 1.15). A floating air bubble formed a contact area with the solid surface of r_1 radius and a boundary film of h thickness. Disjoining pressure in the film was estimated from the capillary pressure of the bubble by measuring the bubble radius and the boundary film thickness: $\Pi = 2\sigma_w/R_0 = \Delta P_c$, where σ_w is the surface water tension, ΔP_c is the capillary pressure of the bubble, and R_0 is the bubble radius. Disjoining pressure in the area of positive values $\Pi(h)$ was estimated at $\sim 10^3 \div 10^4$ dyne/cm^2.

Later, the bubble method was modified, and the films were formed either between two menisci or between a meniscus on a tube end (a capillary) and a plate (Deryaguin and Titievskaya 1953). The measurements based on the same principle conducted by Aronson and Princen (1978) produced the $\Pi(h)$ isotherms for water films on a quartz plate with values $\Pi(h) = 300–1500$ dyne/cm^2.

The methods of disjoining pressure measurements were improved further. This allowed assessing $\Pi(h)$ both in the stable and unstable (dynamic method) areas; the latter was based on the rate of spontaneous thinning at $\partial\Pi/\partial h > 0$ and < 0 (Sheludko 1958). Isotherms of disjoining pressure in capillaries of various radii were obtained by a number of scientists (Deryaguin 1986). It was proven that β-film is not formed in the capillaries of less than the critical radius r_c. Apparently, only α-film may exist in equilibrium with the meniscus at $r_c \leq r$. This allowed scientists to calculate the critical disjoining pressure at 3×10^6 dyne/cm^2, when β-film raptures.

Interesting results were also obtained when the porous filter method was used in combination with the adsorption measurements. Disjoining pressure was measured for film thickness virtually close to the saturation equilibrium ($P/P_s = 0.99$–0.995). For α-film, disjoining pressure was estimated at $\Pi(h) \approx 10^6$ dyne/cm^2 (Deryaguin 1986). Thus, a nearly complete isotherm of disjoining pressure in water solution films was experimentally determinded.

Fig. 1.15 The bubble method for obtaining isotherms of disjoining pressure in the hydrate film of bound water (Deryaguin and Churaev 1984): *1* solid body, *2* bound water, *3* free water, *4* air bubble

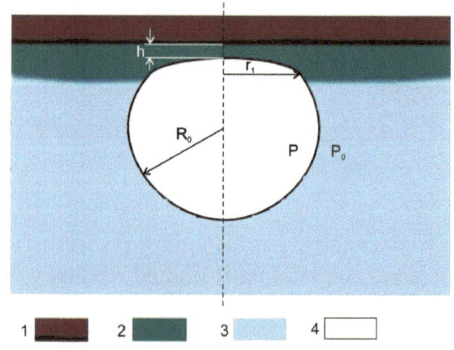

1.3.4.1 Disjoining Pressure of β-film

Deryaguin and Churaev (1984) have summarized the experimental data on the study of the disjoining effect produced by β-films obtained by various researchers (Deryaguin and Kusakov 1936; Deryaguin 1952; Read and Kitchiner 1969; Blake and Kitchener 1972; Aronson and Princen 1978). Figure 1.16 shows the results of this analysis for wetting water films and water solutions of symmetrical electrolytes formed on silicate surfaces.

The obtained data show that the disjoining effect of a water film for silicate surfaces varies, depending on its thickness, from 0.1×10^4 to 1.4×10^4 dyne/cm^2. Under favorable physicochemical conditions, the maximal disjoining pressure of β-film (at its critical thickness) may reach 3×10^6 dyne/cm^2 (Deryaguin and Churaev 1984).

Unlike α-film, the disjoining effect and stability of β-film are not controlled by any specific mechanical properties of the film itself. Its thickness is determined by the forces of ionic-electrostatic repulsion of surface diffusion layers and is controlled by the disjoining effect of these forces. Temperature rise to 65 °C does not cause any structural effects and the β-film thickness remains almost the same.

When the β-film thickness is <60 nm, it raptures. Thinning and destruction of β-film depends on the electrolyte concentration in the solution. The concentration influence is manifested in the decreasing potential of the solid phase and diffusive layers thickness. In addition to the electrolyte concentration, the β-film stability is controlled by a number of other physicochemical factors, such as a decrease in the solution pH to the isoelectric state of mineral surface (Deryaguin and Landau 1941).

Adsorption of ion surfactants exerts a strong effect on the β-film stability. Adsorption of cation-active surfactants on the mineral surface leads to attraction of particles (Aronson and Princen 1978). In this case, the electrostatic constituent of disjoining pressure acquires negative value and leads to a complete disappearance

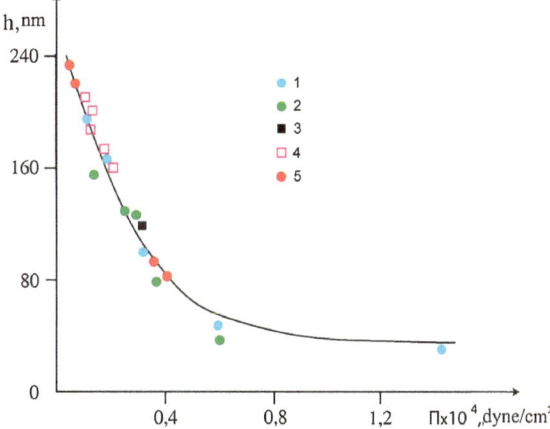

Fig. 1.16 The disjoining pressure isotherm of water β-film at the surfaces of glass, quartz, and mica: *1* water on mica; *2* water on melt quartz; *3* 2×10^{-5} N KCl solution on quartz; *4* 1×10^{-4} N KCl solution on glass; *5* water on quartz (Deryaguin 1986)

of β-branch of the disjoining pressure isotherm. Only α-branch of the isotherm remains stable in this case. Adsorption of anion-active surfactants adds to the negative charge on the mineral surface, which leads to thickening of β-film and increase of its stability.

1.3.4.2 Disjoining Pressure of α-film

The thickness and stability of α-film is controlled by the crystallochemical properties of mineral surface, as well as by specific mechanical and thermodynamic properties of the polymolecular film, composed by adsorbed water molecules, itself. The structural Π_s and molecular Π_m components of disjoining pressure are essential in the disjoining effect of α-films.

Theoretical calculation of disjoining pressure of these films is complicated; therefore, most data have been obtained experimentally by water vapor absorption on flat surfaces of glass, quartz, and mica. The measurements were performed by a number of researchers (Deryaguin and Zorin 1955; Ershova et al. 1979; Pashley and Kitchener 1979) and were summarized by Deryaguin and Churaev (1984) (Fig. 1.17).

A rapid change in disjoining pressure with film thinning is an important feature of α-film. Thus, $d\Pi_s/dh$ for curve 3 (Fig. 1.17) at $h = 5$ nm is equal to 2×10^{17} dyne/cm². This value of the derivative can be explained by neither molecular nor electrostatic forces. According to calculations, for $h = 5$ nm, this forces may produce $d\Pi_s/dh$ no higher than 5×10^{12} dyne/cm², which is four orders of magnitude lower than the experimental data (Deryaguin and Churaev 1984).

Another feature of α-film is the strong dependence of its thickness and disjoining pressure on hydrophilic properties of the surface. Increasing hydrophilic

Fig. 1.17 Experimental values of the adsorbed water α-film thickness at different partial pressures of water vapor (*P/Ps*): *1* for purified surface of quartz; *2, 3* for hydrophilized (slightly purified) surfaces of quartz (Pashley and Kitchener 1979)

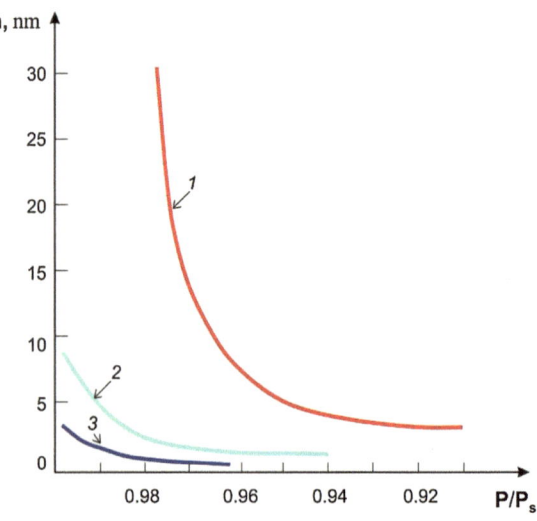

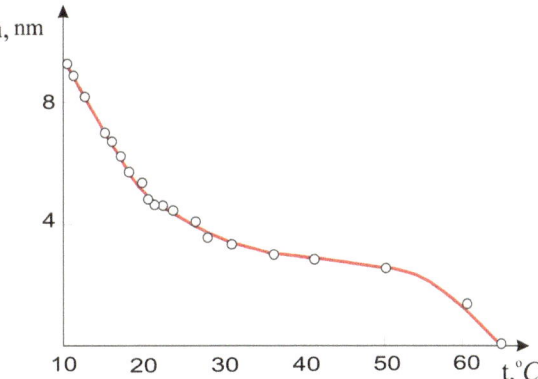

Fig. 1.18 Temperature dependence of the water α-film thickness on quartz (P/Ps = 1)

properties of the surface lead to a rapidly increasing range of action of structural forces. This phenomenon can be explained by the origin of specifically hydrophilic properties of silicate surfaces related to epitaxy processes (Osipov 2011).

Contamination (hydrophobization) of the surface produces an opposite result. On a thoroughly purified quartz surface, α-film is 20–30 nm thick (Pashley and Kitchener 1979). At the same time, even the most insignificant contamination of the surface results in the film thinning to 3–5 nm.

The data on the temperature dependence of the α-film thickness prove directly the structural origin of their stability (Perevertaev and Metsik 1966; Ershova et al. 1975) (Fig. 1.18). A sharp decrease in the α-film thickness upon the temperature rise to 65 °C may be explained by the destruction of the specific structure of boundary layers due to the growing heat motion of water molecules.

1.3.4.3 Disjoining Pressure of Films on Convex and Concave Surfaces

The isotherms of film disjoining pressure discussed above are valid for flat surfaces. It is important to also consider isotherms for uneven surfaces and to analyze the disjoining effect of hydrate films on the surface of spherical particles, filamentous minerals, and faces and edges of clay particles, which may be conventionally assumed as convex surfaces with a certain curvature radii, as well as on concave surfaces inside capillaries.

For a curved surface, the distribution of tangential and normal pressure-tensor components in the film varies, which affects the film stability and thickness. The dependence between the surface liquid tension σ_w and its surface curvature is based on the same effect (Deryaguin and Churaev 1984). The conditions of film stability at a convex surface are described by the following equation (Starov and Churaev 1978):

$$\frac{d\Pi}{dh} < -\frac{m\sigma_w}{(r+h)^2},\qquad(1.49)$$

Fig. 1.19 Variation in the disjoining pressure isotherms at different surfaces: *1* for flat; *2* for convex (Deryaguin and Churaev 1984)

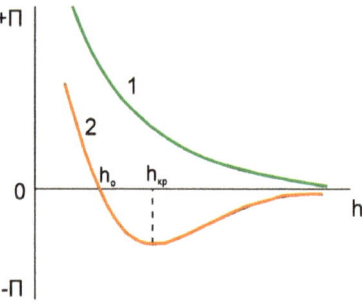

where $m = 1$ or 2 for a film on cylinder or sphere, respectively; σ_w is the surface tension, r is the curvature radius, and h is film thickness.

Analysis of the data obtained suggests that films are less stable at convex than flat surfaces, all other conditions being the same (Fig. 1.19). Comparison of the disjoining pressure isotherms for complete wetting conditions proves that a film of any thickness is stable at a flat surface; whereas, for a convex surface, it loses stability at h_{cr}. For example, for the Hamaker constant $A_0 = -3 \times 10^{-19}$ J and $\sigma_w = 0.023$ H/m, a film on a quartz fiber ($m = 1$) has $h_{cr} = 170, 56$, and 18.5 nm at the surface curvature radius $r = 10^{-3}$, 10^{-4}, and 10^{-5} cm, respectively. α-film exhibits the same effect: a decreasing surface curvature radius leads to thinning of α-film for the same p/p_s (Deryaguin and Churaev 1984). This phenomenon plays an important role in structure formation and in development of face-to edge and edge-to-edge contacts.

References

Aronson MP, Princen HM (1978) Aqueous films on silica in the presence of cationic surfactants. Colloid Polym Sci 256(2):140–149

Babak VG, Kozub SP, Sokolov VN, Osipov VI (1977) Procedure of precision measurement of interaction energy in condensed bodies under different physicochemical conditions. Izv AN SSSR Ser Phys 41:2401–2407 (in Russian)

Babak VG, Sokolov VN, Sveshnikova EV, Osipov VI (1984) Direct measurement of cohesion forces between mica particles in water solutions of polyvinyl alcohol. Inzhenernaya Geologiya (Eng Geol) 4:57–63 (in Russian)

Berezkina GM, Tsareva AM (1968) Transformation of kaolinite clay texture during filtration. In: Modern methods of investigation of physico-mechanical properties of soils and rocks, vol 7, pp 48–52 (in Russian)

Blake TD, Kitchener JA (1972) Stability of aqueous films on methylated quartz. J Chem Soc Faraday Trans Part I 68(10):1435–1448

Churaev NV, Yashenko NE (1973) Rate of water evaporation from capillaries of various diameter. Pochvovendenie (Soil Sci) 9:105–112 (in Russian)

Churaev NV, Sobolev VD, Zorin ZM (1971) Measurement of viscosity of liquids in quartz capillaries. In: Thin liquid films and boundary layers. Academic Press, New York, pp 213–220

Dashko RE (1987) Soil and rock mechanics. Nedra, Moscow (in Russian)

Deryaguin BV (1952) Solvate layers as special boundary phases. In: Proceedings of all-union conference in colloid. Chemistry. Kiev, AN USSR, pp 26–51 (in Russian)

Deryaguin BV (1955) Definition of disjoining pressure. Colloid J 17(3):207–214 (in Russian)

Deryaguin BV (1986) Theory of stability of colloids and thin films. Nauka, Moscow (in Russian)

Deryaguin BV, Abrikosova II (1951) Direct measurement of molecular attraction as a function of distance between surfaces. ZhETF 21(8):1755–1770 (in Russian)

Deryaguin BV, Churaev NV (1972) The isotherm of disjoining pressure of water films on quartz surface. Dokl AN USSR 207(3):572–575 (in Russian)

Deryaguin BV, Churaev NV (1984) Wetting films. Nauka, Moscow (in Russian)

Deryaguin BV, Kusakov MM (1936) Properties of thin liquid layers. Izv AN USSR Ser Chem 5:741–753 (in Russian)

Deryaguin BV, Landau LD (1941) Theory of the stability of strongly charged lyophobic soils. Acta Phys Chem USSR 14(6):633–662

Deryaguin BV, Titievskaya AS (1953) Disjoining pressure of free liquid films. Dokl AN USSR 89(6):1041–1044 (in Russian)

Deryaguin BV, Zorin ZM (1955) The study of surface condensation and absorption of vapors. J Phys Chem 29(10):1755–1770 (in Russian)

Deryaguin BV, Ershova IG, Churaev NV (1972) Stability of wetting films and its influence on liquid evaporation from capillaries. In: Surface forces in thin films and disperse systems. Nauka, Moscow, pp 155–160 (in Russian)

Deryaguin BV, Churaev NV, Muller VM (1985) Surface forces. Nauka, Moscow (in Russian)

Ershova GF, Zorin ZM, Churaev NV (1975) Temperature dependence of polymolecular adsorbed water layer thickness on the quartz surface. Colloid J 37(1):208–210 (in Russian)

Ershova GF, Zorin ZM, Novikova AV, Churaev NV (1979) The study of polymolecular water films on quartz surface. In: Surface forces in thin films and disperse systems. Nauka, Moscow, pp 168–173 (in Russian)

Israelachvili JN (1978) Measurement of forces between two mica surfaces in aqueous solution. J Chem Soc Faraday Trans 74(10):975–1001

Israelachvili JN, Adams LEJ (1978) Measurement of forces in aqueous electrolyte solution in the range 0-100 nm. J Chem Soc Faraday Trans 78(4):975–1001

Israelachvili JN, Pashley RM (1983) Molecular layering of water at surfaces and origin of repulsive hydration forces. Nature 306(5940):249–250

Marčelja S, Radič N (1976) Repulsion of interfaces due to boundary water. Chem Phys Lett 42(1):129–134

Mironenko VA, Shestakov VM (1974) Fundamentals of hydrogeomechanics. Nedra, Moscow (in Russian)

Nikolaev AV (1999) Aspects in human-induced relief of tectonic stress and reduction of seismic hazard. Geoekologiya (Environ Geosci) 5:387–398 (in Russian)

Osipov VI (1973) Mechanism of physicochemical dispersion and stabilization of clay suspensions. In: Issues of engineering geology and soil and rock engineering, Issue 3. Moscow (in Russian)

Osipov VI (1976) Crystallochemical regularities in hydrophilic properties of clay minerals. Herald Moscow Univ Ser Geol 5:107–110 (in Russian)

Osipov VI (2001) Natural disasters at the turn of the XXI century. Geoekologiya (Environ Geosci) 4:293–309 (in Russian)

Osipov VI (2011) Adsorbed water nanofilms in clay, mechanism of their formation and properties. Geoekologiya (Environ Geosci) 4:291–305 (in Russian)

Osipov VI (2014) Physicochemical theory of effective stresses in soils. Water Resour 41(7):801–818

Osipov VI, Sergeev EM (1972) Crystallochemistry of clay minerals and their properties. Bull IAEG 5:9–15

Osipov VI, Sokolov VN (2013) Clays and their properties. GEOS, Moscow (in Russian)

Pashley RM, Israelachvili JN (1984) Molecular layering of water in their films between mica surfaces and its relation to hydration forces. J Colloid Interface Sci 101:501–523

Pashley RM, Kitchener JA (1979) Surface forces in adsorbed multilayers of water on quartz. J Colloid Interface Sci 71(3):491–503

Perevertaev VD, Metsik MS (1966) Investigation of water vapor adsorption on mica crystal surface. Colloidal J 28(2):254–259 (in Russian)

Pereverzeva IN, Osipov VI (1975) Redirection of fabric-forming units of clay dispersions upon a long-term water filtration. In: Physicochemical mechanics and lyophilic properties of disperse systems, Issue 7. Naukova dumka, Kiev (in Russian)

Rabinovich Y (1977) Direct measurements of disjoining pressure in electrolyte solutions as a function of distance between crossed threads. Colloid J 39(6):1094–1100 (in Russian)

Ravina J, Low PF (1972) Relation between swelling, water properties, and b-dimension in montmorillonite-water systems. Clay Clay Minerals 20:109–123

Read AD, Kitchiner JA (1969) Wetting films of silica. J Colloid Interface Sci 30(3):391–398

Shchukin ED, Yusupov RK, Amelina EA, Rebinder PA (1969) Experimental studies of cohesion forces in individual microscopic contacts between crystallites upon pressing together and baking. Colloid J 31(6):913–918 (in Russian)

Shchukin ED, Pertsev AV, Amelina EA (2007) Colloidal chemistry, 5th edn. Vysshaya shkola, Moscow (in Russian)

Sheludko A (1958) Spontaneous thinning of double-sided thin liquid films. Dokl AN USSR 123 (6):1074–1078 (in Russian)

Shofield RK, Samson HR (1953) The deflocculation of kaolinite suspensions and the accompanying change-over from positive to negative charge. Clay Minerals Bull 2:45–51

Starov VM, Churaev NV (1978) Thickness and stability of liquid films on non-flat surfaces. Colloid J 40(5):909–914 (in Russian)

Ukhov SB, Semenov VV, Znamenskii VV, Ter-Martirosyan ZG, Chernyshov SN (1994) Soil mechanics and foundation engineering. ACB, Moscow (in Russian)

Viani BE, Low PF, Roth CB (1983) Direct measurement of the relation between interlayer force and interlayer distance in the swelling of montmorillonite. J Colloid Interface Sci 96:229–244

Zorin ZM, Sobolev VD, Churaev NV (1972) Variation of capillary pressure, surface tension, and liquid viscosity in quartz microcapillaries. In: Surface forces in thin films and disperse systems. Nauka, Moscow (in Russian)

Chapter 2
The Terzaghi Theory of Effective Stress

The theory of effective stress was developed by Terzaghi in the early 1920s (Terzaghi 1925). Its principal positions on stress in soils imply the following: (a) in the water-saturated soil with open porosity, effective stresses represent the excessive stress in the soil skeleton over the neutral stress; (b) effective stresses control the strain-and-stress state, volume variation, and strength independently of neutral stress.

A large body of experimental data supports the validity of these conclusions. Terzaghi theory was successfully used for solving problems in consolidation of porous permeable soils, for explaining reasons of sand liquefaction during earthquakes, as well as for a number of other tasks.

At the same time, the practical experience indicates that the Terzaghi theory cannot be adequately applied to low-permeable fine-grained soils (clay) with closed porosity, as it leads to the discrepancy between calculated and experimental data. The main areas of concerns are the following:

1. The Terzaghi theory uses the total average stresses (σ) produced by external forces; it does consider internal forces creating additional stresses in the soil skeleton. The latter are becoming significantly more important with the finer grain-size of soil.
2. At the same height of the water column, the pore pressure (u) measured in a clay system may differ from the pore (hydrostatic) pressure in a permeable porous body (u_0).
3. Effective stresses are transmitted to the skeleton through the contacts between structural units. Even if the total effective stress is constant, the effective contact stresses may differ due to the disjoining effect of hydrate films, changing amount of contacts, orientation of contact sites, and area of contacts. All these factors influence the strength and deformational properties of soils.

V.I. Osipov, *Physicochemical Theory of Effective Stress in Soils*, SpringerBriefs in Earth Sciences, DOI 10.1007/978-3-319-20639-4_2

Skempton was one of the first scientists who noted unreliable assessment of effective stresses in the Terzaghi theory (Skempton 1960). He introduced the correction for the area of contacts (a_c) to the equation of effective stress:

$$\sigma' = \sigma - (1 - a_c)u, \tag{2.1}$$

where a_c is the ratio between the area of contacts and the total loaded area.

However, like Terzaghi, he also did not take into account the effect of physicochemical forces on effective stresses. Therefore, the introduced correction allowed updating the (σ') values only for some soils, cements, and hard rocks.

Mitchell and Soga analyzed in detail the nature of the effective stresses with the consideration of physicochemical phenomena in clay systems (Mitchell and Soga 2005).

They scrutinized the main physicochemical interactions in the contact zone, such as electromagnetic attraction (Van der Waals forces), electrostatic repulsion and attraction, and chemical forces. Most important, the authors analyzed the stresses in soils at the level of intergrain interaction. For this purpose, they composed the equation of forces acting in soil, including the forces of physicochemical origin:

$$\sigma a + Aa + A'a_c = ua + Ca_c, \tag{2.2}$$

where σ is the external stresses transmitted to the contact; a is the total area of contact where the chemical and physicochemical forces are present; a_c is the area of actual (immediate) contact between particles; A is the far-distance molecular Van der Waals and electrostatic attraction forces; A' is the close-distance attraction forces responsible for the chemical bond and cementation; u is the far-distance electrostatic forces of repulsion equal to $u(a - a_c)$; since $a \gg a_c$, the repulsion may be assumed to be ua; and C stands for the close-distance Born and hydration repulsion forces.

From Eq. (2.2), we may find the stress transmitted to the unit horizontal area equal to the total area of unit contact:

$$\sigma = (C + A')a_c/a + u - A. \tag{2.3}$$

Expression $(C + A')a_c/a$ represents the total stress acting at the true contact in realtion to the total contact area. If we denote the intergrain stress (σ_i') obtained by relating the stress at the true contact to the total contact area (from (2.3)) by $(C + A')a_c/a$, then:

$$\sigma_i' = \sigma + A - u. \tag{2.4}$$

In a similar way, the interparticle effective stresses may be calculated in the partially water-saturated soil. For this purpose, it is necessary to know the pressure in the pore water (u_w) and in the pore air (u_a), as well as the shares, per particle, of areas occupied by water (a_w) and air (a_a):

$$a_w + a_a = a(a_w + a_a)$$

The resulting formula is:

$$\sigma_i' = \sigma + A - u_a - \frac{a_w}{a}(u_w - u_a). \tag{2.5}$$

The equation obtained (2.5) looks similar to that derived earlier by (Bishop 1960):

$$\sigma_i' = \sigma - u_a + \chi(u_a - u_w), \tag{2.6}$$

where $\chi = a_w/a$.

These calculations prove that effective stresses at the contacts can be estimated taking into consideration close- and far-distance physicochemical forces. However, the following questions remain unresolved, specifically, how to: (1) find A and u values in Eq. (2.4), (2) estimate the number of contacts and their area, and (3) transition from assessment of the inter-particle effective stresses σ_i' to assessment of the actual effective stress in soil and rock (σ'').

References

Bishop AW (1960) The principle off effective stress. Norwegian Geotechnical Institute, vol 32, pp 1–5

Mitchell JK, Soga K (2005) Fundaments of soil behavior, 3rd edn. Wiley, New York

Skempton AW (1960) Significance of Terzaghi's concept of effective stress. In: Bjerrum L, Casagrante A, Peck R, Skempton AW (eds) From theory to practice in soil mechanics. Wiley, New York

Terzaghi K (1925) Erdbaumechanik auf Bodenphysilischer Grundlage. Franz Deuticke, Liepzig-Vienna

Chapter 3
Physicochemical Theory of Effective Stresses

The above-discussed results of the Terzaghi theory modification are of great importance for the modern soil mechanics, because they improve understanding of the relationships between the effective stresses in soils and both external and internal physicochemical factors. The data available now support its further development by using the achievements in molecular physics and physicochemical mechanics. Specifically, it is feasible to incorporate the theory of contact interactions developed by Rebinder and his followers, as well as the Deryaguin theory on disjoining pressure of the boundary hydrate films in stress assessment. The chapters below present the author's insight into the subject of effective stresses in soils with the consideration given to these theoretical fundamental positions.

3.1 Types of Contacts in Clays

Transmission and transformation of stresses in porous dispersed systems do not occur through the entire interphase surface, but only in places of the maximal proximity to each other of their components, i.e., at contact sites. The number and type of individual contacts is the most important structural feature controlling the value and type of effective stress transfer.

Depending on the shape of interacting particles, contacts may be of several geometric types, i.e., contacts between spherical particles (Fig. 3.1a), contacts between spherical and flat particles (Fig. 3.1b), and contacts between flat particles (Fig. 3.1c–e). The latter are subdivided into edge-to-edge, face-to-edge, and face-to-face types of contacts.

Despite the large diversity of geometric types, chemical origin, size, and shapes of structural elements forming the porous bodies, there are three main energy types of contacts in disperse cohesive systems: coagulation, transition (point),

© The Author(s) 2015
V.I. Osipov, *Physicochemical Theory of Effective Stress in Soils*,
SpringerBriefs in Earth Sciences, DOI 10.1007/978-3-319-20639-4_3

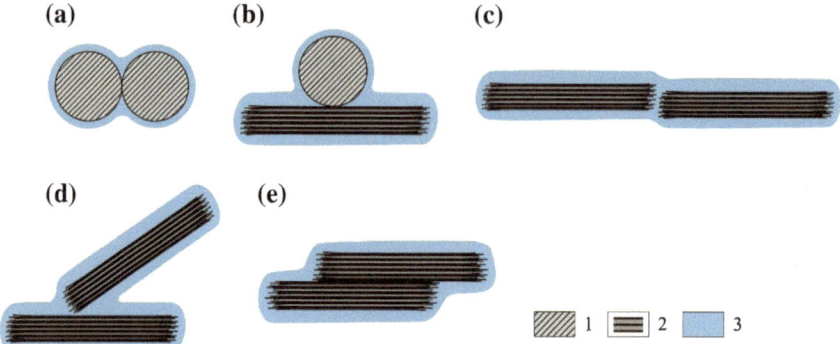

Fig. 3.1 Geometric types of contacts: **a** between spherical particles; **b** between spherical and flat particles; **c–e** between flat particles with the formation of end-to-end (**c**), face-to-end (**d**) face-to-face (**e**) contacts. *1, 2* particles, *3* bound water

Fig. 3.2 Energy types of contacts in dispersed systems: **a** coagulation, **b** transition (point), **c** phase; *1* a mineral particle, *2* bound water

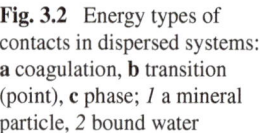

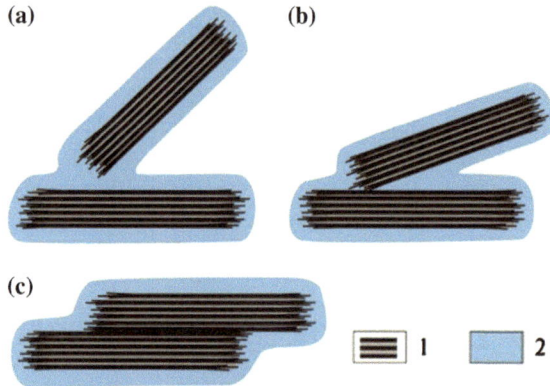

and phase contacts (Rebinder 1958, 1966; Shchukin et al. 2007; Yaminskii et al. 1982). These energy types of contacts are shown for two interacting clay particles in Fig. 3.2. Each type of contact is characterized by a certain mechanism of formation, origin of forces in the contact zone, and the specific features of effective stress transmission.

Coagulation contacts are formed between the interacting particles if the stable film of bound water at the contact is maintained (Rebinder 1958, 1966; Yaminskii et al. 1982) (Fig. 3.2a). In the natural setting, coagulation contacts are found in water-saturated non-compacted or weakly compacted clayey systems, e.g., mud, poorly lithified clayey deposits, and swelled clay of plastic or fluid consistency.

Surface hydrate films that prevent the contact between particles play the most important role in the development of coagulation contacts. Thermodynamically stable state at coagulation contacts is achieved at a certain thickness of hydrate films corresponding to far- or near-potential minimums of the disjoining pressure isotherm (Fig. 1.14). Accordingly, coagulation contacts are usually subdivided into *far-field coagulation* (corresponding to the far-potential minimum) and *near-field coagulation* contacts (corresponding to the near-potential minimum).

Transition contacts form in the course of clay compaction in water-saturated state or with their dewatering on drying. Rapture (destruction) of boundary hydrate films may take place in some rugged (bulged) spots of contacting surfaces under the impact of external and internal factors. Within a small area of direct contact between particles (Fig. 3.2b), the so-called point contacts (the term suggested by Rebinder) are formed via "cold melting" of particles or the formation of ion-electrostatic bonds, similar to those between the layers in clay mineral lattices (Osipov and Sokolov 1974).

The behavior of these contacts is determined by both disjoining effect of hydrate films (in the *coagulation* contact zones) and the stronger chemical and ion-electrostatic forces (in the zones of direct contact of particles); therefore, we call thus formed contact a *transition contact* (Sokolov 1973; Osipov and Sokolov 1974, 1985; Osipov 1979; Osipov and Sokolov 2013).

It is important to assess the effective pressure at the contact area, upon which "cold melting" of particles takes place. This problem may be solved by considering the Hertz contact-problem conditions. The average value of the normal constituent of external pressure in the film-deterioration zone of a hydrate adsorption film is calculated as follows (Yaminskii et al. 1982):

$$P_c = \frac{2}{\pi} \left[\frac{2E^2 f_c}{9(1 - \mu^2)^2 R} \right]^{1/3}, \qquad (3.1)$$

where E—is the Young modulus for particles, μ is the Poisson coefficient, f_c is the efficient external stress at a single contact, and R is the particle radius.

The reference values $E \approx 7 \times 10^{10}$ Pa and $\eta = 0.2$ may be used in calculations. The f_c value is found from the experimental data obtained by Yusupov (1973) in the studies of contact strength between two crystalline bodies under different compression forces. For 1–2 µm-size particles, the threshold compression force causing an abrupt transition from coagulation contacts to stronger contacts is 65×10^{-4} N. Substituting these values into the above-given equation, we identify the critical stress value $P_c \approx 6 \times 10^7$ Pa, upon which the transition contacts between particles start forming.

The strongest *phase contacts* are formed in the course of cementation or with the rising geostatic pressure and temperature (Fig. 3.2c). This type of contacts is based on the ion-electrostatic and chemical forces that affect clay particles and transform them into coarser ultra- and micro-aggregates—crystallites of clay. The formed contacts are similar to the grain boundaries in a polycrystalline body.

Phase contacts exhibit high strength sometimes exceeding the strength of the contacting minerals. They are present in cemented clay deposits, argillites, siltstones, and clayey shales, which are the typical solid bodies manifesting the elastic deformation under external loads and brittle failure upon reaching the limit strength. Unlike transition contacts, phase contacts are practically nonhydratable, which provides for water resistance and the absence of swellability in these rocks.

All these contacts form in the course of geological history. Their formation depends on external factors, above all, on gravitational pressure of the overlying ground mass, and physicochemical conditions. The water-saturation of deposits also plays an important role. Decreasing water saturation leads to conditions that promote transformation of coagulation contacts into transition (point) contacts. Therefore, transition contacts (and even phase [cementation] contacts, in the hot arid climate) may be formed in soils that have been in the aeration zone for prolonged time.

The lithification degree of clay is the key factor for contacts' formation in such deposits. With the growing lithification degree, coagulation contacts (typical of mud and poorly lithified plastic clay) gradually become transition contacts (compacted clay and semi-solid consistence). This is observed when external load from the overlying deposits is higher than 10–30 MPa. At stresses equal to 80–200 MPa, the formed phase contacts are typical of cemented clay, argillite, and aleurolite (Osipov and Sokolov 2013).

Depending on the type of contacts, the role of external and internal factors in the development of effective stress varies. Physicochemical factors exert the strongest effect on the behavior of soils with coagulation contacts. Their role is lesser in clay with transition contacts, and it becomes insignificant in soils and rocks with phase contacts. Therefore, below we analyze the influence of physicochemical factors on the effective stress in soils with the coagulation and transition types of contacts.

3.2 Number of Contacts

The effective contact stresses are calculated in the contact areas normal to the direction of external stress. In order to calculate normal efficient stresses with the consideration of internal stresses, it is necessary to know the number of contacts where the internal and external stresses are transmitted and the total value of the true contact stress.

For the assessment of the number of contacts, let us consider an arbitrary horizontal surface of area (s) inside a porous body, which is perpendicular to the applied external stress. Actually, the surface is slightly undulating as shown in Fig. 3.3.

Estimation of the number of contacts per unit area of horizontal surface is based on the assumption that all individual contact areas between structural elements are also normally directed to the exerted stress.

At present, several calculation schemes, i.e., models for dispersed porous bodies, have been developed to estimate the number of contacts per unit area of horizontal surface. The number of contacts in these models is controlled by the size and shape of particles and the type of their packing, which is closely related to

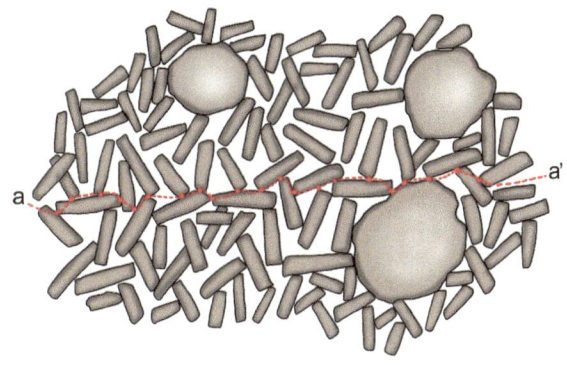

Fig. 3.3 Horizontal surface: a–a′ cross sectional line in a porous body

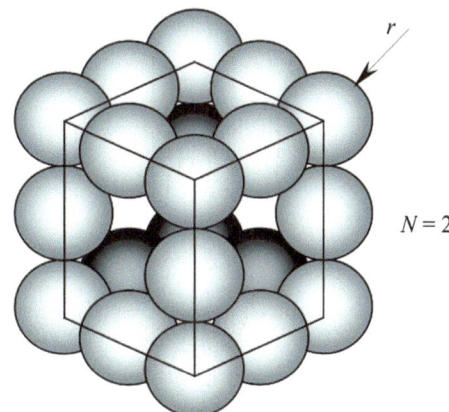

Fig. 3.4 Schematic representation of the globular model for disperse porous structures: N average number of particles from node to node depending on structures' porosity; r average radius of particles

$N = 2$

porosity. The lesser the particle size and the denser the packing, the more contacts occur per unit area of horizontal surface.

The simplest model is *the globular model* suggested by Rebinder et al. (1964) for the structures composed of spherical particles with porosity >48 %. Later, this model was expanded to include the structures with porosity 26–48 % (Amelina and Shchukin 1970). The globular model assumes the straight chains of spheres of equal diameter touching each other (Fig. 3.4). The chains stretch in three dimensions; they cross and form structural nodes. The packing type is described by a structural parameter N equal to the average number of particles between two nodes. For $N = 1$, the lattice with a simple cubic packing is formed, and for a fractional value of N, the system has irregularly alternating nodes. In this model, porosity n is related unequivocally to parameter N. Graphically this relationship is expressed by the function (Fig. 3.5) (Amelina and Shchukin 1970):

$$\frac{1}{N^2} = f(n) \tag{3.2}$$

Fig. 3.5 Dependence between $1/N^2$ parameter and porosity

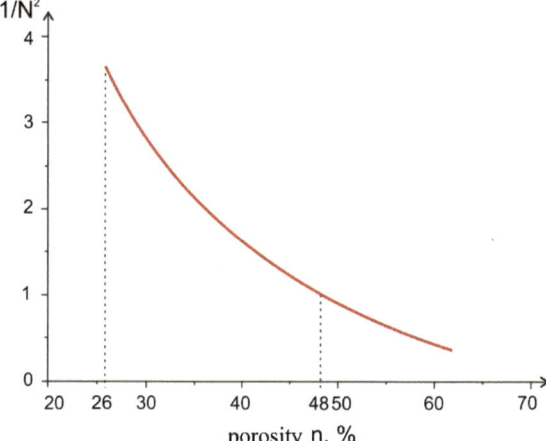

The number of contacts per unit area of horizontal surface (χ) is related to parameter N and the average radius of structural element r according to the relationship (Rebinder et al. 1964; Shchukin 1965; Amelina and Shchukin 1970):

$$\chi = \frac{1}{4r^2N^2} \tag{3.3}$$

The globular model may be applied to sand, sandstone, siltstone, and some fine grained deposits with the structural elements (particles, micro-aggregates) close to a spherical shape. For example, it may be applied for calculating the number of contacts in certain sandy loams, loams composed of sandy grains and round-shaped sandy-clayey aggregates, as well as in tripolith and diatomaceous soils. The calculations prove that the number of contacts in silica clay is within $10^5 \div 10^7$ cm^{-2}, depending on its porosity; while it is $1.2 \times 10^6 \div 6.2 \times 10^7$ cm^{-2} in loessial loam.

The main drawback of the globular model is that it does not take into account size diversity and shape anisotropy of the structural elements composing clay. For polydisperse systems, Sokolov (1991) proposed a *bidisperse globular model*, which allows assessing the number of contacts in the system composed of coarse (R radius) and fine (r radius) particles (Fig. 3.6). According to this model, the total number of contacts is equal to the number of contacts between coarse particles (χ_R), multiplied by the number of contacts between fine particles (χ_r), occurring within the contact area between coarse particles. The χ_R and χ_r values are found from the following formulas, respectively:

$$\chi_R = \frac{3z(1-n)}{8\pi R^2} \tag{3.4}$$

$$\chi_r = \frac{\rho_R \varphi_r R^2}{2\rho_r \varphi_R r^2}, \tag{3.5}$$

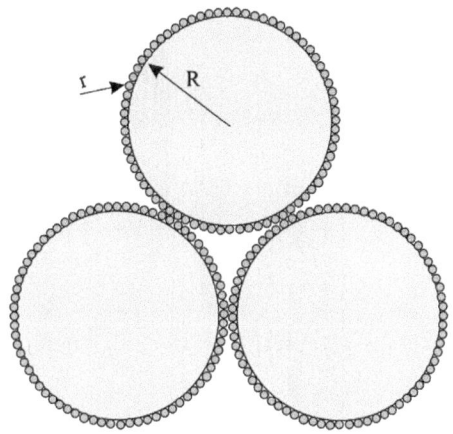

Fig. 3.6 Schematic rerpesentation of bidisperse model of porous structure: *r* radius of fine particles and *R* average radius of coarse particles (Rebinder et al. 1964; Sokolov 1991)

where ρ_R, ρ_r, φ_R, φ_r, R u r are density, content, and average equivalent radius of coarse and fine particles, respectively; z is the coordination number; and n is porosity.

The total number of contacts in a bidisperse system is:

$$\chi = \chi_R \chi_r = \frac{3z(1-n)\rho_R\varphi_r}{16\pi r^2 \rho_r \varphi_R} \qquad (3.6)$$

Calculations performed for loam according to (3.6) prove that value χ may range from 1.2×10^7 cm^{-2} to 1×10^9 cm^{-2} depending on the ratio between φ_R and φ_r for the loam porosity of 40 %.

To describe the clay deposits composed of flat anisometric particles, a *"twisting card-house"* model was developed (Sokolov 1985, 1991). The model is based on the idea that thin discs of platy clay particles or foliaceous microaggegates touching each other represent the "walls" of this "card-house" (Fig. 3.7).

For estimating the compaction degree of soils, the suggested model uses parameter θ, which stands for the average angle between particles. The porosity is related to θ value via formula:

$$n = 1 - \frac{K \cdot b/a}{sin\theta + (K \cdot b/a)}, \qquad (3.7)$$

where K is the coefficient (for discs, $K = 3\ \pi/4$); a and b are the average diameters and thicknesses of discs, respectively.

The considered model allows assessing the number of contacts per unit area of the contact failure surface (χ^θ) as a function of the average angle between the clay particles (θ):

$$\chi^\theta = \chi^{90}/sin\theta. \qquad (3.8)$$

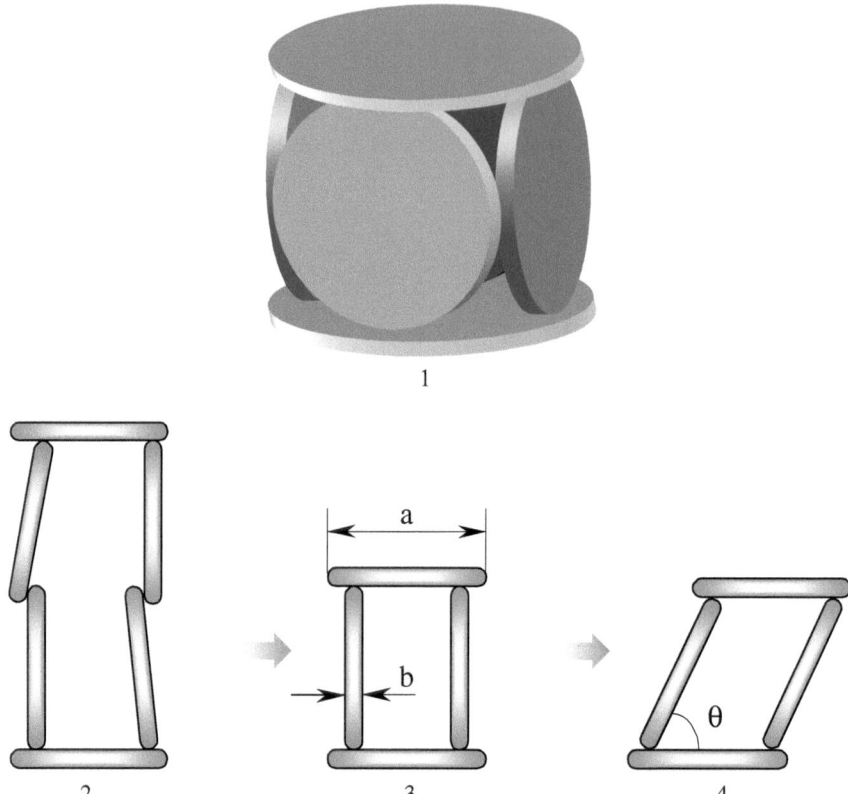

Fig. 3.7 Schematic representation of "twisting card-house": **1** general view; **2, 3, 4** different stages of structure transformation on compaction, where a is the particle length, b is the particle thickness, and θ is the angle of the particle's inclination (Sokolov 1991)

The number of contacts χ^{90}, when the cell-forming particles are perpendicular to each other ($\theta = 90°$), is described as $\chi^{90} = 2/a^2$, where a is the particle diameter.

Thus, Eq. (3.8) may be written as:

$$\chi^{\theta} = 2/sin\theta \cdot a^2. \tag{3.9}$$

Parameters a and θ can be estimated through quantitative evaluation of clay microfabric images obtained by scanning electron microscope.

Calculations performed according to (3.9) show that for the modern Caspian mud, the χ^{θ} value is 6.9×10^7 cm^{-2}. At the same time, for marine clay of the Khvalynian age collected near Volgograd, with practically similar grain size as the mud but with a higher orientation degree of the structural units, this value is 3.9×10^8 см$^{-2}$. The χ^{θ} value increases to 4.4×10^8 см$^{-2}$ for clayey shale. This rather moderate increase of the χ^{θ} value is associated with the fact that despite the minimal θ values the size of microaggregates in shale (an increasing a parameter in Eq. (3.9)) is higher.

3.3 Area of Contacts

3.3.1 Coagulation Contacts

The area of coagulation contact between two spherical particles ($a_c = a_{c(c)}$) depends on their radii (R) and the hydrate film thickness (h) (Fig. 3.8a). Parameter (a_c) characterizes the overlapping zone formed by hydrate films of contacting particles, where disjoining pressure exists. For two spherical particles, the unit contact area is found from the equation:

$$a_c(c) = 2\pi \, R\left(h - \frac{1}{2}\right), \tag{3.10}$$

where h is the hydrate film thickness on the particle surface outside the contact, l is the minimal thickness of the contact hydrate layer, and R is the particle radius.

Calculations based on Eq. (3.10) show that the contact area between two similar spherical particles is:

(1) for $R = 0.5$ μm (500 nm), $h = 50$ nm, $l = 50$ nm, $a_c \approx 8 \times 10^4$ nm$^2 \approx 8 \times 10^{-10}$ cm^2;

(2) for $R = 50$ μm (5×10^4 nm), $h = 50$ nm, $l = 50$ nm, $a_c \approx 8 \times 10^6$ nm$^2 \approx 8 \times 10^{-8}$ cm^2.

The contact area between two flat particles of the same size positioned at an angle θ (Fig. 3.8a) is assessed differently:

$$a_{c(c)} = \left[\min\left(\frac{h}{Sin\frac{\theta}{2}} - \frac{l}{Sin\theta}, a\right) + \min\left(\frac{h}{Sin\frac{90°-\theta}{2}} - \frac{l}{Sin(90° - \theta)}, H + 2h\right)\right] \cdot b \tag{3.11}$$

where a and b are the width and length of two interacting flat particles, respectively; H is the particle thickness; h is the hydrate film thickness outside the contact, and l is the minimal thickness of hydrate film in the contact area.

Expressions

$$\min\left(\frac{h}{Sin\frac{\theta}{2}} - \frac{l}{Sin\,\theta}, a\right) \text{ and } \min\left(\frac{h}{Sin\frac{90^0-\theta}{2}} - \frac{l}{Sin(90^0 - \theta)}, H = 2h\right)$$

mean that the lowest of the considered values are assumed, that is, the following condition is met:

min (a, b) = a, if a ≤ b, or min (a, b) = b, if a ≥ b.

Clearly, for $\theta \to 0$, that is, the particles are parallel and overlap each other completely, the contact area is controlled by a and b dimensions of the overlapping particles. For the particles with $a = b = 1$ μm, the a_c value is 10^6 nm^2 or 10^{-8} cm^2. For $\theta \to 90^0$, the particles are perpendicular to each other. The contact area is minimal in this case and is equal to 1.6×10^5 nm^2 (1.6×10^{-9} cm^2) for the same particles.

The area of coagulation contact between a spherical particle and a flat particle can be found from the equation:

$$a_{c(c)} \approx 2\pi\, R(h - l)$$

(3.12)

The contact area between a spherical particle of R radius and a flat particle twice exceeds the contact area between two spherical particles of the same radius as in the latter case.

3.3.2 Transition Contacts

The total area of a transition contact a_c consists of (1) the total area of the sites of immediate tangency ($a_{c(s)}$) and (2) the total area of the coagulation sites ($a_{c(c)}$) (Fig. 3.8b). Assessment of the area of the first group, i.e., ("dry") sites (where the hydrate film has been squeezed out) appears to be most difficult.

The minimal and maximal areas of "dry" sites in a transition contact area may be estimated from the tensile strength of these contacts and their hydration

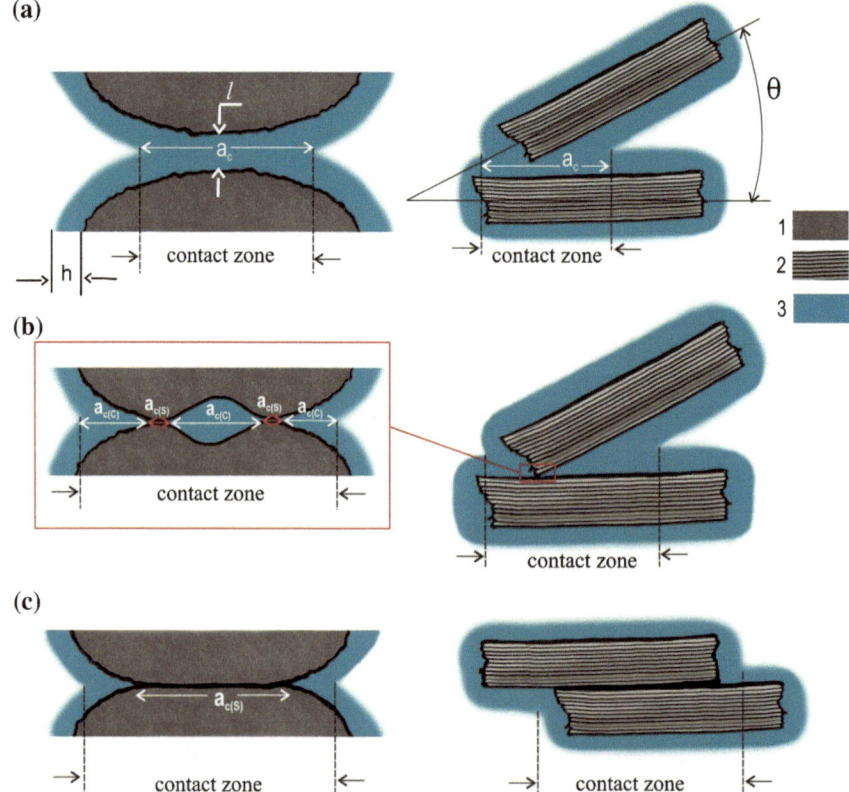

Fig. 3.8 Area of coagulation (**a**), transition (**b**), and phase (**c**) contacts: *1, 2* particles, *3* hydrate film of bound water

capacity. For this purpose, the most important feature of transition contacts can be used, specifically, their metastability or ability to transition into coagulation contacts under certain conditions. This phenomenon takes place when external pressure is relieved and clay receives additional wetting. When these conditions are met, disjoining pressure at the coagulation sites of a transition contact exceeds the tensile strength of the chemical bonds formed at the sites of the immediate contact between the particles. Due to the loss of chemical bonds, the "dry" sites are hydrated and, thus, coagulation contacts appear.

Experiments prove (Osipov and Sokolov 1974) that a transition contact between two micron-size particles loses its reversibility at a strength of $\sim 3 \times 10^{-7}$N. At a higher strength, transition contacts behave as irreversible phase contacts. The value 3×10^{-7} N may be apparently taken as the upper boundary of the transition contact strength. As was already noted, transition contacts are formed through chemical (valence) bonds. Assuming a single valence bond equal to $e^2/4\pi\varepsilon_0 b^2 \approx 10^{-9}$ N, (where e is the electron charge, ε_0 is the dielectric constant equal to 8.85×10^{-12} C^2/N m^2, and b is the distance between atoms in a crystalline cell ~ 0.1 nm) it has been calculated that for the formation of a transition contact with the maximal strength of 3×10^{-7} N, 3×10^2 bonds are needed. When the number of chemical bonds is $<3 \times 10^2$, the formed transition contact is metastable. Taking into account the crystalline fabrics of faces in phyllosilicates, it can be calculated that 3×10^2 bonds can be formed within an area of 35 nm^2.

For a known $a_{c(s)}$ value and by calculating the total contact area (a_c) from one of the Eqs. (3.10–3.12), it is possible to determine a share of "dry" and coagulation areas in relation to the total contact area ($[a_{c(s)}/a_c]$ for the coagulation area and $[1 - a_{c(s)}/a_c]$ for the contact area).

3.4 Effective Contact Stress in Water-Saturated Soils

3.4.1 Coagulation Contacts

Thin hydrate interphase films at the coagulation contacts exert a substantial impact on stress distribution at these contacts. As was mentioned earlier (Sect. 1.3), disjoining pressure of stable hydrate films at the contact area, i.e., $\Pi(h)$ value, is an integral result of different surface forces present at the contact (molecular, electrostatic, and structural-mechanical). The product of disjoining pressure and the contact area (a_c) equals the total value of inner physicochemical forces acting at a coagulation contact:

$$\sigma_c^{in} = \Pi(h) \cdot a_c. \tag{3.13}$$

Besides the internal forces, there are the effective stresses produced by general external impacts (equal to σ'/χ, where χ is the number of contacts per unit area of contact surface) are transmitted to each contact.

The actual effective stress value per unit coagulation contact equals to the difference between the external and internal forces:

$$\sigma_c' = \sigma'/\chi - \Pi(h)a_c \tag{3.14}$$

3.4.2 Transition Contacts

As was mentioned above, the transition (point) contacts consist of a series of small sites of immediate ("dry") contiguity of particles at the convex parts of contacting surfaces. At the same time, the remaining part of contact between bulging areas retains the features of coagulation contact and, thus, the boundary hydrate films (Fig. 3.8b).

There is no interphase hydrate layer at the sites of immediate contact of particles ("dry" sites), therefore, effective stress at these sites is transmitted directly to the soil skeleton. At the coagulation sites, disjoining pressure of hydrate film is present; it reduces the external effective stress at these sites. The actual effective stress at transition contacts is the integral result of all stresses transmitted to "dry" and coagulation contact areas.

Effective stresses transmitted to the sites with direct contact of particles ($\sigma'_{c(s)}$), may be calculated from the following equation:

$$\sigma_{c(s)}' = \frac{\sigma' a_{c(s)}}{\chi a_c}, \tag{3.15}$$

where σ' is the total external effective stress, $a_{c(s)}$ is an area of "dry" sites at the transition contact, a_c is the total area of transition contact, and χ is the number of contacts per unit area of horizontal surface.

Similarly, the effective stresses may be estimated at the coagulation sites of contact ($\sigma'_{c(c)}$):

$$\sigma_{c(c)}' = \left(\frac{\sigma'}{\chi} - \Pi(h)a_{c(c)} \right) \times \left(1 - \frac{a_{c(s)}}{a_c} \right), \tag{3.16}$$

where $\Pi(h)$ is disjoining pressure of the hydrate film.

Thus, the total effective stress at the transition contact is:

$$\sigma_c' = \sigma_{c(s)}' \sigma_{c(c)}' + \frac{\sigma'}{\chi} - \Pi(h) \cdot a_{c(c)} \left(1 - \frac{a_{c(s)}}{a_c} \right), \tag{3.17}$$

where $a_{c(c)}$ is the area of coagulation sites at the transition contact.

3.4.3 Phase Contacts

As was discussed earlier, the phase contacts receive the entire effective stresses and transmit them to the soil skeleton (Fig. 3.8c). The influence of physicochemical factors on the distribution of effective stresses is negligible there. Hence, the Terzaghi law describes adequately coarse fragmentary and most sandy soils, as well as fractured hard rock formations including cemented clay deposits such as argillites, siltstones, and clay shales.

3.5 Effective Contact Stresses in Unsaturated Soils

Effective stresses appearing at the contacts of unsaturated soils have specific features related to the formation of capillary menisci and the capillary pressure. The formed menisci and capillary water create additional internal stresses at the contacts (may be found from expressions (1.29) and (1.31)). The capillary pressure appears under capillary condensation at moisture content close to the maximal air-dry moisture (W_{mg}) and disappears at moisture content close to the full saturation of soil.

As was mentioned earlier, the mechanism of action of capillary forces depends on the degree to which pores are filled with water. For soils in the aeration zone above the capillary fringe, capillary menisci are formed due to capillary condensation of the air moisture (Fig. 1.8). Capillary water occurs in soil below the capillary fringe (Fig. 1.9). In the former case, the action of capillary forces is manifested through the pressure of capillary menisci on contacts; in the latter case, through the pressure of capillary water transmitted to the entire soil skeleton. Therefore, the capillary pressure effect should be considered separately for these two cases.

Clay soils in the aeration zone above the capillary fringe are usually characterized by transition contacts. Considering this fact and using Eq. (1.29), the value of effective stresses at an individual contact, accounting for the capillary menisci effect, is:

$$\sigma_c' = \frac{\sigma'}{\chi} - \Pi(h)\, a_{c(c)} \left(1 - \frac{a_{c(s)}}{a_c}\right) + \pi r \sigma_w \left(1 - \frac{r_1}{r_2}\right), \qquad (3.18)$$

where r_1 and r_2 are the menisci radii in the contact spacing and σ_w is the surface tension of water.

The effective contact stresses acting in clay below the capillary fringe are calculated taking into account the pressure of the water mass (m_w) capillary lifted to the height h_c on the soil skeleton. This pressure is equal to $h_c m_w g$. Then, for deposits with prevailing coagulation contacts, the value of contact effective stresses, considering the capillary forces, is:

$$\sigma_c' = \frac{\sigma'}{\chi} - \Pi(h)a_{c(c)} + \frac{h_c m_w g}{\chi}, \qquad (3.19)$$

where g is gravity acceleration.

If transition contacts prevail in the deposit, the equation is modified:

$$\sigma_c' = \frac{\sigma'}{\chi} - \Pi(h)\, a_{c(c)}\left(1 - \frac{a_{c(s)}}{a_c}\right) + \frac{h_c m_w g}{\chi} \tag{3.20}$$

3.6 Actual Total Effective Stresses in Soils

Actual total effective stresses in clay soil, developing under the impact of external (gravitational) and internal (physicochemical) impacts, may be calculated by using the effective contact stresses obtained above, the number of contacts per unit area of a soil cross-section, and the equation for pore pressure in open and closed porous systems. Depending on the prevailing types of contacts and the degree of soil water-saturation, the calculation equations differ.

3.6.1 Water-Saturated Soils

Coagulation contacts. Mud and weakly lithified clay with coagulation contacts and open pore network manifest the effective contact stress found from Eq. (3.14), and the pore pressure equal to u (at the stage of filtrational consolidation) and $u = 0$ (at the stage of secondary consolidation). In this case, the equation for the actual total effective stress in these soils takes the following form:

$$\sigma'' = \chi\left[\frac{\sigma - u}{\chi} - \Pi(h)a_c\right] = (\sigma - u) - \chi\Pi(h)a_c, \tag{3.21}$$

where χ is the number of contacts per unit area of a horizontal area, $\Pi(h)$ is disjoining pressure of hydrate films, a_c is the contact area, σ is the total external stress, and u is pore pressure.

Clay soils either with a high content of clay minerals (of smectite composition, in particular) or subjected to a certain compaction in the course of lithogenesis (but retaining its coagulation structure) manifest a somewhat different kind of stress. This is related to the formation of closed-pore space structure. In some isolated pores of this structure, pore pressure that remains not lower than the pressure of the initial filtration gradient in this clay (u'), may form. Hence, the actual total effective stress in this clay may be found from the following expression:

$$\sigma'' = [\sigma - u'] - \chi\Pi(h)a_c \tag{3.22}$$

Transition contacts. The stress state of soils with transition contacts is controlled by the existent areas of direct contact between particles and the coagulation contacts.

For such contacts, stresses can be found from Eq. (3.17). Most of soils with transition contacts have experienced some compaction in the course of lithogenesis. Therefore, they are treated as over-compacted soils in engineering practice. The pore pressure in over-compacted soils may develop only if their natural over-compaction pressure threshold is exceeded, which is most unlikely in engineering practice, because this threshold in water saturated soils with transition contacts is known to be 20–30 MPa. Therefore, the total external stress $\sigma' = \sigma$ should be taken as the effective external stress in these soils. Then, the expression for the actual total effective stress in water-saturated soils with transition contacts is as follows:

$$\sigma'' = \sigma - \chi \Pi(h) a_{c(c)} \left(1 - \frac{a_{c(s)}}{a_c}\right) \tag{3.23}$$

where $a_{c(s)}$ and $a_{c(c)}$ is the area of immediate ("dry") and coagulation contacts, respectively; and a_c is the total contact area.

Phase contacts. Soils and rocks with phase contacts are the cemented or crystalline formations, where the contacts resemble the boundaries between the grains in a polycrystalline body. In the absence of fracturing, effective stresses in these rocks fit the general stress $\sigma' = \sigma$. In water-saturated fractured hard rocks, effective stresses are calculated taking into account the uplift water effect.

3.6.2 Unsaturated Soils

The actual total effective stress in unsaturated clay depends on its moisture content. At moisture content below W_{mmg}, the compression forces of capillary menisci appear. Considering that transition contacts prevail under such conditions, σ'' is:

$$\sigma'' = \sigma - \chi \Pi(h) \left(1 - \frac{a_{c(s)}}{a_c}\right) + \chi \pi r_1 \sigma_w \left(1 - \frac{r_1}{r_2}\right), \tag{3.24}$$

where r_1 and r_2 are the radii of menisci in the contact gap (Fig. 1.9), and σ_w is the surface tension of water.

When moisture and capillary saturation of soil increase, the compression effect of capillary water increases too. The actual total effective stress for such conditions is:

$$\sigma'' = \sigma - \chi \Pi(h) \left(1 - \frac{a_{c(s)}}{a_c}\right) + h_c m_w g, \tag{3.25}$$

where m_w is the mass of capillary water, g is gravity acceleration, and h_c is the height of capillary uplift.

References

Amelina EA, Shchukin ED (1970) The study of some regularities in the formation of contacts in porous dispersed systems. Colloid J 32(6):795–800 (in Russian)

Osipov VI (1979) Origin of strength and deformational properties of clay. Moscow State University, Moscow (in Russian)

Osipov VI, Sokolov VN (1974) Role of ion-electrostatic forces in the formation of structural bonds in clay. Herald of Moscow Univ. Series Geology 1:16–32 (in Russian)

Osipov VI, Sokolov VN (1985) Theory of contact interactions in soils. In: Sergeev EM (ed) Theoretical fundamentals of engineering geology. Physico mechanical fundamentals. Nauka, Moscow (in Russian)

Osipov VI, Sokolov VN (2013) Clays and their properties. GEOS, Moscow (in Russian)

Rebinder PA (1958) Physicochemical mechanics. Znanie, Moscow (in Russian)

Rebinder PA (1966) Physicochemical mechanics of dispersed structures. In: Physicochemical mechanics of dispersed structures. Nauka, Moscow (in Russian)

Rebinder PA, Shchukin ED, Margolis LYa (1964) Mechanical strength of porous dispersed bodies. Dokl.AN USSR 154 (3):695–698 (in Russian)

Shchukin ED (1965) Some problems in physicochemical theory of strength of fine-dispersed porous bodies—catalysts and adsorbents. Kinet Catal 6(4):641–650 (in Russian)

Shchukin ED, Pertsev AV, Amelina EA (2007) Colloidal chemistry, 5th edn. Vysshaya shkola, Moscow (in Russian)

Sokolov VN (1973) The study of formation of structural bonds in clays upon their dehydration. Abstract of Cand.Sci. Dissertation. Moscow (in Russian)

Sokolov VN (1985) Physicochemical aspects of clay mechanics. Eng Geol 4:28–41 (in Russian)

Sokolov VN (1991) Models of clay microstructures. Eng Geol 6:32–40 (in Russian)

Yaminskii VV, Pchelin VA, Amelina EA, Shchukin ED (1982) Coagulation contacts in dispersed systems. Khimiya, Moscow (in Russian)

Yusupov RK (1973) Physicochemical studies of formation conditions and strength of microscopic phase contacts between individual solid particles. Abstract of Cand.Sci. Dissertation. MSU, Moscow (in Russian)

Conclusion

The discussion presented above proves that stresses in soils and rocks are a complex integral result of external impacts and internal processes, extending far beyond a simple consideration of the applied load and pore pressure as was suggested earlier by Terzaghi. The structural properties of soils, their composition, and the degrees of lithification and water saturation are of great importance for the assessment of stress state. Contacts between structural elements are the main areas, on micro-scale, where external and internal stresses are redistributed. Transformation of contact types, their number, and area over the course of deformation is the principal factor controlling clay soil behavior.

The analysis of the entire array of factors allows us to conclude that understanding of the laws of clay behavior under different loads and different physicochemical conditions requires deep knowledge of the internal structure of soil and the processes involved in its response to these impacts and conditions. Therefore, further success in soil mechanics, in many respects, will be determined by the progress in learning the nature of processes existent in soils. This requires a transition from the phenomenological approach to gaining insight into the essential mechanisms present in soils. The physicochemical soil mechanics represents the foundation for this transition.

© The Author(s) 2015   55
V.I. Osipov, *Physicochemical Theory of Effective Stress in Soils*,
SpringerBriefs in Earth Sciences, DOI 10.1007/978-3-319-20639-4